The Evolution of In...

Need to understand sec. 21

The Evolution of Insect Flight

Andrei K. Brodsky

Department of Entomology
St Petersburg University

Oxford   New York   Tokyo
OXFORD UNIVERSITY PRESS

Oxford University Press, Great Clarendon Street, Oxford OX2 6DP
Oxford New York
Athens Auckland Bangkok Bombay
Calcutta Cape Town Dar es Salaam Delhi
Florence Hong Kong Istanbul Karachi
Kuala Lumpur Madras Madrid Melbourne
Mexico City Nairobi Paris Singapore
Taipei Tokyo Toronto
and associated companies in
Berlin Ibadan

Oxford is a trade mark of Oxford University Press

Published in the United States by
Oxford University Press Inc., New York

First published 1994
First published in paperback 1996

A catalogue record for this book is available from the British Library

Library of Congress Cataloging in Publication Data
Brodsky, Andrei K.
The evolution of insect flight/Andrei K. Brodsky.—[1st ed.]
Includes bibliographical references (p.   ) and index.
1. Insects—Flight.   2. Insects—Evolution.   I. Title.
QL496.7.B76   1994   595.7′01852—dc20   93–24790
ISBN 0 19 850089 0

Printed in Great Britain by
Bookcraft (Bath) Ltd
Midsomer Norton, Avon

Preface

Insect flight is a subject of great interest to scientists working in many fields. It involves many challenging problems of great complexity, not least because of the many subtle movements made by insects and the great difficulties inherent in observing free-flying animals. The subject may be approached from an anatomical, physiological, morphological, energetic, aerodynamic, or ecological perspective. It may be tempting to try to study only one of these aspects of flight, but this should be discouraged, for they are all interlinked and we can learn a great deal from studying different aspects simultaneously. The most fruitful approach is to study the changes which occurred in the structure and function of insect wing apparatus in the course of evolution.

This book is not for easy reading at idle moments. It is addressed to the interested, thoughtful reader who will not be deterred by the need to master scientific terms and concepts. As a rule, I haven't attempted to simplify described phenomena, but have instead paid special care to explain the more difficult and complex questions, trying to avoid particularly specialized terminology. The material on which the present work is based is derived partly from the literature, and partly from collaborative research with postgraduates and students of the Department of Entomology of St Petersburg University: O.A. Antonova, V.P. Ivanov, V.D. Ivanov, and D.L. Grodnitsky. The author is very grateful to the following entomologists who have shared their speculations on the flight behaviour of different insects: O.M. Bocharova-Messner, L.A. Zhiltzova, M.V. Kozlov, A.L. Lvovsky, C.U. Sinev, A.A. Stekolnikov, and M.I. Phalkovich. I thank Dr H.K. Pfau, Institut für Zoologie, Johannes Gutenberg Universität Mainz, Dr R.J. Wootton, Department of Biological Sciences, University of Exeter, Dr J. Kukalová-Peck, Department of Earth Sciences, Carleton University, and the Canadian Journal of Zoology for permission to reproduce figures from their publications.

Conscious that no translation can satisfy every reader, I can only hope that those unfamiliar with Russian will find the present version of the book acceptable if not congenial, bearing in mind the formidable style of the original. The book was translated by S. Scalon. I am indebted to the artist I.G. Guy for the beautiful illustrations. I am also obliged to D. Kirsanov who prepared the manuscript and made many useful notes about the translation. I would welcome any reader's suggestions and criticisms arising from the book.

A.K.B.

St Petersburg
March 1993

Contents

Symbols and abbreviations

A	stroke angle	S	length of the path taken by the wing tip during a full flap
$a, b,$ and c	main air streams in a wake of flying insect		
B	body angle	Sr	Strouhal number
b	stroke plane angle relative to the horizontal	T	thrust
		t	time
C_D	drag coefficient	U_S	flow velocity in an insect's wake
C_L	lift coefficient	u	component of relative wind due to flapping movement of wing
D	total drag of flying insect		
D'	wing drag	V	flight speed; flow speed
G	body mass	v	component of relative wind due to forward movement of insect
g	gravitational acceleration		
h	gliding height	w	wind on wing
L	total lift	Γ	circulation
L'	lift generated by the wing	α	aerodynamic angle of attack
l	characteristic dimension; wing length; gliding distance	Θ	gliding angle
		λ	aspect ratio (wingspan/chord)
n	wingbeat frequency	μ	viscosity of medium (air); gliding coefficient
R	resultant force = full aerodynamic force		
Re	Reynolds number	ν	kinematic viscosity of medium (air)
S	characteristic area of the body; wing area; cross-sectional area of flow tube	ρ	density of liquid (air)

Abbreviations of structural terms

The following abbreviations of structural terms are used throughout the text, and are based on Snodgrass' (1927, 1929, 1935) and Matsuda's (1970) widely accepted nomenclature.

On the figures, the main wing base hinges are circled; Roman numerals designate segment number; the symbol '&' denotes structures which are continuous with each other and membranous and unsclerotized parts of the skeleton are shaded.

T		tergum
P		pleuron
	EPM	epimeron
	EPS	episternum
	PLA	pleural apophysis
	PLS	pleural suture
	PWP	pleural wing process
S		sternum
	STA	sternal apophysis
L		leg
	CX	coxa
	TN	trochantin
	TR	trochanter

Parts of the tergum

ATG	acrotergite
LPN	lateropostnotum
N	notum
PH	phragma
PN	postnotum
PAA	postalar process (postalar arm)
PRA	prealar process (prealar arm)
PRN	pronotum
PRSC	parascutum
PSC	prescutum
p-SCT	pseudoscutellum
SC	scutum
SCT	scutellum

Sutures, sulci, and ridges of the tergum

AC	alacrista
ACS	antecostal suture
ALSS	anterolateral scutal sulcus
LPS	lateral parapsidal suture
MLS	medial longitudinal suture
OS	oblique suture
PLSS	posterolateral scutal sulcus
PS	parapsidal suture
PSS	supplementary parapsidal suture
PSCS	prescutoscutal suture
RSSS	recurrent scutoscutellar suture
SSS	scutoscutellar suture
TF	tergal fissure
TNS	transnotal suture
TSF	transscutal fissure
TSS	transscutal sulcus
TWG	tergal wing groove

Sclerites of the wing base

ASC	subcostal apodeme
AX	axillary
BA	basalar
BCU	basicubital (basicubitale)
BPCU	basipostcubital (basipostcubitale)
BR	basiradial (basiradiale)
BSC	basisubcostal (basisubcostale)
DMP	distal medial plate
HP	humeral plate
PMP	proximal medial plate
SA	subalar

Wing processes of the notum

ANP	anterior
AMNP	anteromedian
MNP	median
PMNP	posteromedian
PNP	posterior

Veins of the wing

C	costal
Sc	subcostal
R	radial
Rs	radial sector
M	medial
Cu	cubital
CuA	anterocubital
CuP	posterocubital
PCu	postcubital
A	anal

The transverse veins, other than the humeral vein (h), are named after the longitudinal veins which they join, using lower-case letters; for example, cu-pcu.

Wing folds

pb	basal
ph	humeral
psr	radial sector
pm	medial

pcu	cubital
pcua	anterocubital
ppcu	postcubital
pa	anal
pt	transverse
po	oblique
pj	jugal
p3ax	third axillary sclerite
pv	vannal

Wing tracheae

c	costal
sc	subcostal
r	radial
m	medial
cu	cubital
pcu	postcubital
a	anal

Other structures

AP	apodeme
AXC	axillary cord
F	fosse
e	scutellum edge
h1	main hinge of the tergum
h2, 3, . . .	other hinges of the tergum
Li	ligament
s1	superior layer of sclerite
i1	inferior layer of sclerite
fw	forewing
hw	hindwing
TG	tegula

Introduction

If ornithopters are ever to replace aeroplanes, their design will demand even further study of bird and insect flight mechanisms.

K.E. Tzialkowsky.

Insects were the first organisms in the life of the Earth to rise into the air. The existence of wings made insects faster and more manoeuvrable, enabled them to migrate regularly, and greatly complicated their behaviour. Nutritional and reproductive possibilities grew, and new ways of escaping from enemies appeared. The origin of flight was an important evolutionary step, and not only for insects. Species-rich communities, manifold and intimate associations with plants, complex ecological relationships—all these became possible when winged insects took to the air 300 million years ago.

The advantages of flight over any other form of locomotion gave rise to a sudden increase in species diversity and the dispersal of insects all over the world. A broad range of forms appeared. Furthermore, the origin of flight led to radical reconstruction of the whole insect structure with profound changes in musculature, sense organs, reproductive organs, as well as other structures.

Throughout their long period of development, insects' flight mechanisms progressed and their flight capacities were more and more refined. The structure and function of the insect flight apparatus is a question of great interest for comparative physiology, taxonomy, applied entomology, and biomechanics. Such features as wing structure, wing articulation, and the skeleton and musculature of the thorax are widely used for the identification of insect groups, from species to orders. Moreover, familiarity with mechanisms and parameters of flight can be important when planning expensive pest control measures. Certain features of wing structure and the principles involved in generating aerodynamic forces might be used in engineering. Our knowledge of the main principles of the air-wing interaction has to contribute a great deal to refinements and projecting of non-steady propulsive devices.

In 1930, the famous American morphologist and entomologist Snodgrass published his book *How insects fly*. In his clear and concise style, the author examined the structure of the wing apparatus, considered the palaeontological evidence, and discussed the problems of kinematics and aerodynamics—in other words, he covered all the areas he considered to be necessary to answer the question posed by the title of his book. Notwithstanding the simplicity of some of his concepts, Snodgrass appears to have been much closer to the real answers than many investigators who have come after him. Later on there appeared Pringle's two thorough surveys (1957, 1968) and the brilliant research of the German biophysicist Nachtigall, summarized in his monograph *Insect in flight. A glimpse behind the scenes in biophysical research* (1974). Weis-Fogh, the leading expert in insect flight, also published a number of important theoretical works, for example 'Flapping flight and power in birds and insects, conventional and novel mechanisms' (1975).

More than half a century has passed since Snodgrass raised the question of how insects fly. Much new information on the flight of different insects has been obtained, and among those studied are species which until then had never attracted much attention: dragonflies, mayflies, caddisflies, and others. Substantial progress has been made in working out methods for studying insect flight; high-speed cinematography, hot-wire anemometry, and quick-response techniques for measuring forces are very widely used. Visualization of the aerodynamic wake of a flying insect has been achieved for the first time. The progress that has been made in the study of insect flight allows us to come close to building a complete picture of the phenomenon. Data on wing structure interact closely with the concepts of kinematics and wing deformation, and so do the aerodynamics of flight with its ecological and behavioural features.

To analyse the functional relationships between different parts of the wing apparatus and to discover the principles underlying its construction, the flight-providing system is best divided into three subsystems:

(1)   an inner locomotive mechanism;
(2)   the surrounding air;
(3)   the control subsystem.

The forces necessary for flight are generated by the interaction between the wing and the surrounding air. The way the wings are put into motion, the characteristics of this motion, and the main features of the wing-airflow interaction form the essence of the flight mechanics of insects. In this book considerable attention is paid to the mechanics of flight and how they have changed in the course of evolution. To reconstruct the functionally and aerodynamically controlled sequence of changes in wing apparatus of the first winged insects, one needs to know the principles of its organization, laws of kinematics, and how wing deformation occurs in flight; and one certainly needs to have an adequate understanding of the interaction between the flapping wing and the air.

We intend to take a peep into the world where for a third of a billion years, the flight mechanism of insects has been gradually refining itself, and where insect flight skills have developed under the influence of natural selection. Significant changes in wing apparatus organization have occurred only rarely during the evolution of winged insects. Flapping flight itself has only evolved once. At the same time, the loss of the ability to fly, due to wing reduction or otherwise, is a widespread phenomenon in the evolution of the pterygotes. Gliding, the power-saving mode of flight, has arisen many times in the course of evolution. The ability to

glide has evolved at least five times, namely in the Palaeodictyoptera, mayflies, and dragonflies, and also in some Neuroptera and Lepidoptera. By contrast, a transition to flapping flight from gliding has never been documented. Furthermore, instances of the loss of the ability to fold wings are well known, but there is no known instance of this ability being regained. Things are apparently the same with modes of wing folding. The transition from a roof-like arrangement to a flat one has been attended by profound changes in wing articulation. Such changes have been described for cockroaches, some dipterans, and hymenopterans (Brodsky 1989). However, no example of a transition from the flat to the roof-like arrangement of the wings on the back is known, except in aquatic bugs.

All this lends support to the belief that the evolution of the wing apparatus of insects is characterized by a tendency for certain changes to occur more often than others. Thus, flapping flight, the ability to fold the wings over the abdomen, their roof-like arrangement—all of these have arisen only once in the course of evolution. Wing loss, inability to fold wings, the origin of flat folding, and gliding flight have arisen many times and in different groups. When examining the possible paths which the evolution of pterygotes may have taken, and assuming the probability of various changes in the wing apparatus structure, we have to take into account that large body size is closely related to the ability to glide. The inability to fold wings over the back has something to do with the frequent use of gliding flight, although no species capable of gliding has ever been recorded among flies that do not fold their wings at rest. This might be explained by the small body size of these insects. Doubtless, such characteristics of the wing apparatus as the wide vanni of the hindwings and the flat arrangement of the folded fore- and hindwings over the tergum are also related, as is the case for cockroaches. However, the development of new folds also enables the hindwings to be locked easily under the roof-like folded narrow forewings, as it is, for example, in some caddisflies and moths.

The flight characteristics both of individual

species of insects and of their natural group-ings are described in this book. The rapid inflow of new data and breakdown of old concepts prevent us from adopting one uni-versally accepted taxonomy for winged insects. However, Martynov's system is currently more or less generally accepted. Martynov's system of winged insects (1928) was as follows:

Division I. Palaeoptera: Odonata, Ephemer-optera
Division II. Neoptera
  Subdivision Polyneoptera
    1. Blattopteroidea: Blattodea, Mantodea, Isoptera
    2. Orthopteroidea: Orthoptera, Phasmodea, Plecoptera, Embioidea
    3. Dermapteroidea: Dermaptera, Hemimeroidea
    4. Thysanoptera: Thysanoptera
  Subdivision Paraneoptera
    1. Corrodentia: Psocoptera, Zoraptera, Mallophaga, Anoplura
    2. Rhynchota: Homoptera, Heteroptera
  Subdivision Oligoneoptera
    1. Coleoptera, Strepsiptera, Megaloptera, Neuroptera, Raphidioptera, Hymenoptera
    2. Mecoptera, Diptera, Aphaniptera, Trichoptera, Lepidoptera

Later, Martynov (1938) moved the Thysanoptera complex to the subdivision Paraneoptera—a decision appreciated by contemporary students. This classification still stands, without any sig-nificant alterations. The most radical changes to Martynov's system have been proposed by palaeoentomologists. Using rigorously dated fos-sils one can follow the evolution of two sepa-rate branches, the Gryllones and Scarabaeones, from the most primitive winged insects, the Protoptera. The Gryllones went back to living in gaps in the substrate of the forest floor, under-went embryonization of development, gained the ability to fold the wings flat over the abdomen, and eventually gave rise to the sub-division Polyneoptera in Martynov's system. The Scarabaeones retained for a long period the habits

and the morphology of Protoptera, which are believed to feed on sporangia of Gymnospermae, and later gave rise to the ancestors of recent mayflies, and then to the ancestors of recent dragonflies, and finally to the common ancestors of Paraneoptera and Oligoneoptera.

Work on the systematics of insects is still in progress. Contemporary systematics of insects is continually being perfected, and is based on the comparison of different facts and views of those studying them. In this book the following systematics of winged insects is accepted (extinct divisions are marked with a dagger†):

Infraclass Palaeoptera
  Cohort 1
    1. Superorder Palaeodictyopteroidea: Palaeodictyoptera+, Megasecoptera+, Permothemistida (Archodonata)+
    2. Superorder Diaphanopteroidea: Diaphanopterodea+
  Cohort 2
    1. Superorder Geropteroidea: Geroptera+
    2. Superorder Ephemeropteroidea: Permo-plectoptera+, Ephemeroptera
    3. Superorder Odonatoidea: Meganeurida+, Odonata

Infraclass Neoptera
  Cohort Paoliiformes
    1. Superorder Protopteroidea: Protoptera+
  Cohort Polyneoptera
    1. Superorder Plecopteroidea: Plecoptera, Embioptera, Grylloblattida, Dermaptera
    2. Superorder Blattopteroidea: Blattodea, Mantodea, Isoptera
    3. Superorder Orthopteroidea: Protorthoptera+, Titanoptera+, Orthoptera, Phasmodea
  Cohort Paraneoptera
    1. Superorder Caloneuroidea: Caloneurida+, Blattinopseida+
    2. Superorder Hypoperloidea: Hypoperlida+
    3. Superorder Psocopteroidea: Psocoptera
    4. Superorder Thysanopteroidea: Thysanoptera
    5. Superorder Homopteroidea: Homoptera and Heteroptera

Cohort Oligoneoptera
1. Superorder Miomopteroidea:
    Miomoptera (Palaeomanteida)[+]
2. Superorder Coleopteroidea: Coleoptera,
    Strepsiptera, Megaloptera
3. Superorder Neuropteroidea:
    Neuroptera, Raphidioptera
4. Superorder Mecopteroidea: Mecoptera,
    Trichoptera, Lepidoptera, Diptera
5. Superorder Hymenopteroidea:
    Hymenoptera

PART I

Basic principles of insect flight

According to the laws of contemporary aerodynamics the cockchafer shouldn't fly but it flies. If we could manage to determine aerodynamics of flight of the cockchafer we'd either find imperfection in the contemporary theory of insect flight or discover that the cockchafer possesses some unknown way of generating high lift.

Inscription on a laboratory door

The life of an insect, like that of any other living creature, consists of a finite number of behavioural sequences. Each sequence comprises a long series of events, which begins when the insect gets a stimulus from the environment. Transformed into nerve impulses, this stimulus gives rise to a locomotory act: for example, flight muscles contract and distort the top of the thorax and press down on the wing bases causing elevation of the wings. Thus the insect sets out towards the source of the stimulus. It pursues prey, approaches an individual of the opposite sex, performs stereotyped acts to assure its partner of its 'good will', it mates, and so on. All this lasts until another stimulus takes over, and another behavioural sequence begins.

Let us try to break the chain of events and to analyse in detail the parts which are closely related to the work of the wing apparatus and flight. But what is 'flight'? This question may seem strange until we remember that it is not only winged animals that drift in the air, sometimes for long distances. Air currents carry a number of small arthropods: spiders, mites, wingless insects, and others. But winged insects produce aerodynamic forces themselves, by flapping their special flight organs—the wings, and do not depend on the wind. The principal requirement for flight organs is that they should create a propulsive force; by producing it, they transmit a certain amount of energy to the surrounding air. Successful colonization of the air became possible because insects mastered the art of creating thrust. The wings made insects faster and more manoeuvrable, but in addition, some insects have mastered a passive flight regime in which they simply drift about in the air with their wings fixed in an outstretched position. The ability to fly enhanced insect survival, and enabled them to settle all over the world. Finding food, reproducing, and escaping from enemies all became much easier. New, specialized food sources became available. Flying animals gained the ability to use such patchy and ephemeral food sources as carrion, because they were able to travel long distances in search of carcasses.

1

Structure of the wing apparatus

The locomotory centre of winged insects is the thorax. The two wing-bearing segments (the meso- and metathorax) form the 'pterothorax'. It is in the thoracic muscles of the wing that energy is transformed from chemical into mechanical form. Hence, the pterothorax can be compared with the engine of an aeroplane (Fig. 1.1). Power generated by contraction of the muscles is transmitted to the propulsive device—the wings—through the skeleton and the system of the axillary sclerites in the wing bases. The insect flaps its wings and thus generates aerodynamic forces. The insect uses different flight regimes depending on what it is doing.

The insect wing, an outgrowth of the thoracic wall, consists of two cuticular layers tightly connected to each other. The dorsal wing surface grades into the dorsal part of the skeleton, and the ventral surface into the lateral wall of the segment (Fig. 1.2). Sclerotized skeletal elements—the axillary sclerites—occur at the point of articulation. The axillary sclerites have fixed topographic relationships among themselves. Moreover, there are a number of joints in the same area, which is why this area is often called the axillary apparatus.

Insect wings, unlike those of flying vertebrates, lack intrinsic musculature and are activated by contraction of the thoracic muscles. The muscles and the wings are functionally connected by the extrinsic skeleton of the pterothorax and the wing articulation. When contracting, the muscles distort the top of the thorax and thus press down on to the wing bases, causing elevation of the wings (Fig. 1.3). The insect wing acts as a first-order lever, so no force need be applied distally—on the wing blade—for it to be elevated or depressed. It is sufficient for the proximal end

to be depressed or elevated over a small arc, by pressure from the edge of the tergum. Changes in the shape of the tergum are brought about by the 'indirect flight muscles' (Fig. 1.3).

The wing articulates with the tergum at two main points: the anterior, at the first and the second axillary sclerites, and the posterior, at the third axillary sclerite (Fig. 1.2). Inferiorly, the second axillary rests on the pleural wing process, which is an extension of the pleural suture. The basalar sclerite is situated in front of the pleural process, and is connected to the wing via the basisubcostale. The subalar sclerite lies behind the pleural process, and is connected to the wing via the inferior cuticular layer of the second axillary sclerite (Fig. 1.4). The wing articulation complex also includes the median plates, the basal part of the veins, and the folds of the wing. The subalar and basalar sclerites, the processes of the tergum, the third, and sometimes the first axillary sclerites are attached by the 'direct flight muscles' (Fig. 1.4).

Thus there is an intimate relationship between the wing apparatus and its component parts; contraction of the muscles brings about changes in the parts of the skeleton which are either directly or indirectly connected with the wing bases. In addition, the musculature of the pterothorax is multi-purpose: it not only causes the wings to move but it also makes the legs move, regulates the positions of the various parts of the insect body, takes part in respiratory movements, thermoregulation, and may also carry out other functions.

There is a great deal of information available about the structure of the pterothorax in different insects. Much of the credit for our understanding of the generality and the classification of

Aeroplane

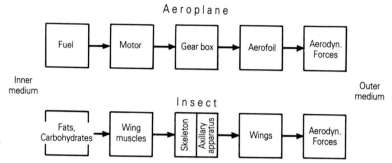

Inner
medium

Outer
medium

Insect

Fig. 1.1 Comparison of insect and aeroplane flight system.

the wing apparatus, and of the structure and function of its components can be attributed to Snodgrass (1927, 1929, 1935). Snodgrass established the nomenclature for the insect's thoracic skeleton, later improved by Matsuda (1970), and their nomenclature has become generally accepted. An attempted revision by Dicke and Howe (1978) was unsuccessful. With some modifications, the purpose of which will be explained later, we use Snodgrass' and Matsuda's nomenclature throughout this book (see Abbreviations for structural terms, pp. xiii–xiv).

1.1 Pterothorax structure

Each wing-bearing segment is made up of a dorsal (tergum), ventral (sternum), and a couple of lateral parts (pleura). The extent of development of the wing-bearing segments varies among insect species. The Odonata characteristically have extremely small terga. Since such terga are too small to act as intermediaries between the indirect wing muscle-depressors and the wing bases, the downstroke is caused exclusively by contraction of the direct flight muscles. The terga of other extant insects are characterized by relatively large plates with deeply indented margins, and bear various sutures, grooves, and ridges.

The pleural suture, which extends between the articular processes of the wing and the leg, provides a fulcrum for the wing. The pleural suture divides the pleuron into two parts, the episternum and the epimeron. The pleural apophysis projects from the medial part of the suture. The pleural apophysis is usually associated with the corresponding sternal apophysis, which arises from the point of articulation of the coxa and the sternum (Fig. 1.2). These two apophyses, which are joined by muscle fibres, form the internal skeletal structure that strengthens the segment

wall. Correspondingly, another wall is furnished with the same endoskeleton.

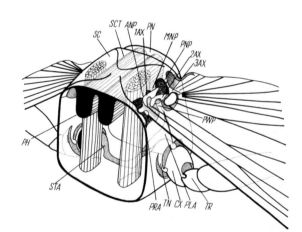

Fig. 1.2 Transverse cross-section of the wing-bearing segment, showing leg and wing bases, the wing articulation region and endoskeleton. The pleural and dorsal regions are drawn as though transparent.

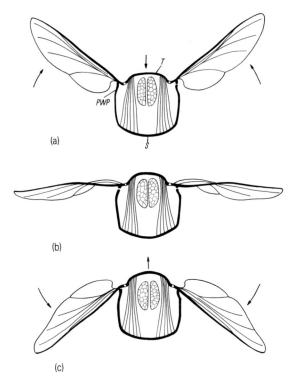

(a)

(b)

(c)

Fig. 1.3 Cross-section of the insect thorax illustrating the action of the indirect wing muscles. (a) Contraction of the dorsoventral muscles causes the tergum to be depressed and the wings to move upward (*T*, tergum; *S*, sternum; *PWP*, pleural wing process). (b) Unstable wing position. (c) Contraction of the dorsal longitudinal muscles causes the tergum to arch upward and the wings to move downwards (after Snodgrass 1935).

1.1.1 The organization of a generalized tergum

The tergum plays an important role in the production of wing movement by responding to the contraction of the indirect flight muscles. To examine the form and direction of the changes which have taken place in the terga of the different phylogenetic branches, we need to identify the main features of the generalized tergum including all the most common features which occur among the various insect orders. Most previous attempts to establish a generalized wing-bearing tergal plate structure have

tended to follow the structure of a particular insect group, such as the Orthoptera (Snodgrass 1935) or the Megaloptera (Shvanvich 1949). Such a scheme might be considered a generalized one if it could easily be translated into the terga of other orders of insects. Clearly, the fewer transformations that are needed, the greater the success of the generalization. Such a generalized scheme must express the underlying characteristics of the terga of many phylogenetically distant groups of insects. The generalized tergum shown in Fig.1.5 satisfies the requirements mentioned above. It can be transformed into a typical mayfly tergum (using 6 single transformations, such as 'opening' the scutoscutellar suture, removing muscle *t20*, and others); a stonefly (4 transformations), a cockroach (5), an orthopteran (4), a cicada or a bug (5), one of the Oligoneoptera (5), and so on. Even the highly specialized terga of dragonflies can be derived from this scheme. At the same time it should be pointed out that the generalized scheme of the tergum cannot be considered as the phylogenetically primitive pattern, since it includes, where possible, specialized features which are typical of extant orders.

The wing-bearing tergal plate is subdivided by sutures of various origins into an acrotergite,

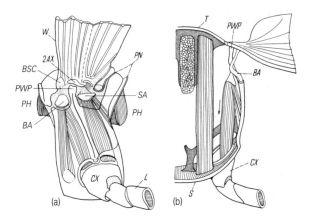

Fig. 1.4 The insect thorax, illustrating the action of the direct wing muscles. (a) Lateral view of the thorax showing basalar and subalar muscles. The pleural region is drawn as if transparent. (b) Cross-section of the thorax showing the mode of action of the basalar muscles (after Snodgrass 1935).

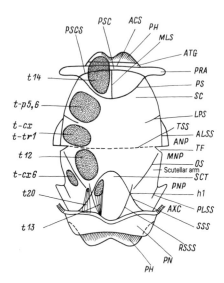

Fig. 1.5 The principal features of a generalized wing-bearing tergal plate. The sites of attachment of the different groups of indirect flight muscles are shaded on the left-hand side.

notum, and postnotum; and the notum can in turn be divided into prescutum, scutum, and (for most insects) scutellum. The system of sutures and sulci of the tergum is highly evolutionarily labile, so that even slight modifications of tergal plate movement during a wing stroke may cause marked structural changes. To understand how the tergum is modified during evolution, it is important to determine the original function of each tergal structure. The sutures and sulci of the tergum fall into three functionally distinct groups:

(1)   sutures inherited from wingless ancestors, whose original functions are obscure;

(2)   sutures and sulci which facilitate the deformation of the tergal plate during the wing-stroke;

(3)   sutures which separate the sites of attachment of the indirect wing muscles on the inner surface of the tergum.

The first group includes the antecostal and prescutoscutal sutures. Matsuda (1970) believed that these were inherited from wingless ancestors, and were therefore homologous with the

corresponding sutures of *Lepisma*. In winged insects the antecostal suture demarcates the acrotergite posteriorly, and ventrally forms the phragma, which is produced by inflection of the ectoderm. The prescutoscutal suture separates the prescutum from the scutum, which is situated posteriorly; the lateral processes of the prescutum are called prealar arms.

The sutures and sulci of the second and third groups result from the acquisition of wings, and are related to the work of the wing mechanism. Unlike the sutures, the sulci of the tergum are formed by linear thinnings of the skeleton.

Three sulci—the transscutal, the anterolateral scutal, and the posterolateral scutal—and two sutures—the recurrent scutoscutellar and the oblique—form the second group. The anterolateral and the posterolateral scutal sulci respectively divide the anterior and the posterior angles of the scutum, which bear the notal wing processes. Both sulci provide lines of 'weakness' which increase the flexibility of the corresponding parts of the scutum. The transscutal sulcus joins the left and right tergal fissures, enabling the tergum to bend when the dorsal longitudinal muscles contract. The transscutal sulcus is obviously present in both segments of dragonflies, in the metathorax of some Orthoptera, and in the Hymenoptera. In other cases (in mayflies, cicadas, bugs, etc.) this term is currently used to describe the suture which projects obliquely from the intersection of the scutoscutellar and the recurrent scutoscutellar sutures, and which diverges towards the tergal fissure on each side of the scutum. Frequently, this 'oblique suture', whose name reflects its position on the scutum, divides the convex medial part of the scutum from its lateral flat parts (Fig. 1.5). This oblique suture is apparent in all winged insects except those where the structure of the wing-bearing tergal plate is completely altered, as in cockroaches.

The last suture of the second group is the recurrent scutoscutellar suture. It is well developed in Ephemeroptera, Plecoptera, Orthoptera, etc.; that is, in relatively primitive insects. Evolution of this suture allowed the notum to deform during contraction of the dorsal longitudinal muscles. The oblique sutures and the recurrent scutoscutellar suture together form stiffening ribs. When the dorsal longitudinal muscles are contracted, and the anterior and the posterior phragmas thereby pulled together, the stiffening ribs elevate the lateral parts of the scutum around the anterior and the median notal wing processes. The flat parts of the scutum which lie lateral to the oblique sutures form the arms of the scutellum. Each scutellar arm bears on its apex a median notal wing process (Fig. 1.5). The external margin of the scutellar arm carries the posterior notal wing process, separated by the posterolateral scutal sulcus.

Two pairs of sutures—the parapsidal and the lateral parapsidal—and two single sutures—the medial longitudinal and the scutoscutellar—form the third group. The development of these sutures was associated with differentiation of the indirect wing muscles' areas of attachment on the notum. Each suture is matched by a ridge on the inner notum surface which gives rigidity to the area in which the muscle is attached. The ridges and their corresponding sutures on the outer notum surface are made prominent by the two muscles which are inserted on each side, which is why the pair of parapsidal sutures and the scutoscutellar suture are the most conspicuous of all the tergal sutures. The dorsal longitudinal and the tergopleural muscles are attached to both sides of the parapsidal sutures, while the dorsal oblique and the scutellar muscles are attached to the scutoscutellar suture. Sometimes sutures may appear as concave linear flexures of the sclerotized part of the notum, or as membranous lines which are closed at each end. The lateral parapsidal sutures of mayflies provide an example of the former type, while the parapsidal sutures of some cicadas are an example of the latter.

1.1.2 The musculature

Each skeletal muscle of an arthropod forms an intimate connection between two areas of external cuticle, one of which remains fixed during muscular contraction, while the other area is movable. The movable end of the muscle is usually slender and is connected to the sclerite via an apodeme, an inflection of the internal skeleton. The other end of the muscle is broader, and provides a fulcrum for the active end. Sometimes both ends of the muscle are morphologically similar, making it difficult to identify their relative functions, which is why muscle classification is widely based on topographical principles.

Matsuda (1970) therefore grouped muscles according to their attachment points. Hence the muscles are classified in the following topographical groups: tergal (t), pleural (p), tergopleural (t–p), tergosternal (t–s), tergotrochanteral (t–tr), tergocoxal (t–cx), pleurocoxal (p–cx), pleurotrochanteral muscles (p–tr), and pleurosternal muscles (p–s). The pleurosternal muscles are represented in each wing-bearing segment by the pair of muscles p–$s1$ which connect the opposite ends of the pleural and sternal apophyses on each side of the segment and regulate the flexibility of the segment wall.

All the muscles which end inferiorly on the leg are bifunctional: they move the leg while walking and move the wing while flying. They comprise pairs of antagonistic muscles which immobilize the legs during the work of the wings and vice versa. Wilson (1968*b*) showed experimentally that the anterior tergocoxal muscle t–$cx2$ in Orthoptera appears to be the levator of the wing and the promotor of the coxa, that the posterior tergocoxal muscles t–$cx6$ and t–$cx7$ are the levators of the wing and the remotors of the coxa, that the subalar muscle t–$cx8$ is the supinator and the depressor of the wing and the remotor of the coxa, and that the basalar muscle p–$cx2$ is the pronator and depressor of the wing, and promotor of the coxa (Fig. 1.6). This arrangement applies to all insect species which have a complete set of the principal wing muscles.

Even though the functions of individual pterothorax muscles are difficult to establish, these muscles can nevertheless be divided into their principal functional groups. The first step in this direction is the division of the muscles into two groups—indirect muscles and direct muscles. The muscles of the first group are attached to the tergum either by their superior end or by both ends.

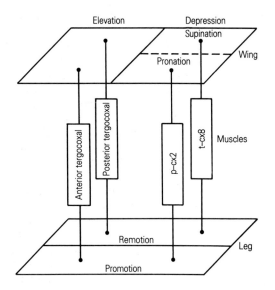

Fig. 1.6 Mode of action of muscle-antagonists in the insect pterothorax (after Wilson 1968b).

The indirect wing muscles include tergal, tergopleural, tergocoxal, and tergotrochanteral groups (see Fig. 1.5, p.10). The tergocoxal muscles are subdivided into anterior muscles (*t–cx1*, *t–cx2, t–cx3*) and posterior muscles (*t–cx6, t–cx7*); the former are inserted on the anterior coxal margin, and the latter on the posterior margin. The upper end of muscle *t–cx1* appears to have the most anterior position on the scutum, and is often attached to the lobe of the prescutum. This muscle is characteristic of insects with short transversal terga (cockroaches and some orthopterans). The next tergocoxal muscle (*t–cx2*) is present in nearly all insects, with the exception of mayflies. The *t–cx3* muscle is less frequently present; it has been described in cockroaches, mantids, and mayflies, but while cockroaches and mantids have both *t–cx2* and *t–cx3*, mayflies only have *t–cx3* (although it may strictly be *t–cx2*).

The upper end of one of the posterior tergocoxal muscles (*t–cx6*) is inserted on the posterior part of the scutum, near the dorsal oblique muscle (Fig. 1.5); the upper end of the other one (*t–cx7*) lies on the anterior part of the scutum, near the anterior tergocoxal muscles. In mayflies the upper ends of muscle *t–cx7* are attached in the area between the anterior and posterior wing processes (Fig. 1.7a), which is presumed to be its original position. One of the two posterior tergocoxal muscles, *t–cx6*, is reduced in Mecopteroidea. In caddisflies the dorsal oblique muscle is weak and, because of this, the upper end of *t–cx7* has migrated caudally towards the usual position of *t–cx6*,

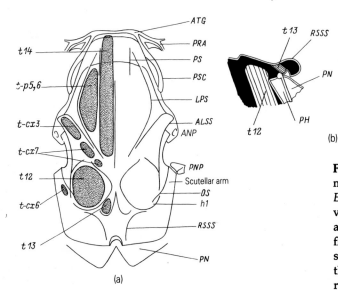

(a)

(b)

Fig. 1.7 Tergum of the mesothorax of the mayfly *Ephemera vulgata*. (a) Dorsal view showing the sites of attachment of the indirect flight muscles. (b) Sagittal section of the posterior part of the notum, with the muscle *t14* removed.

which is absent. In cicadas and bugs only one posterior tergocoxal muscle is present, but it is difficult to tell exactly which it is. However, as in cicadas the upper end of the posterior tergocoxal muscle lies medially relative to the oblique suture, while in bugs it lies laterally, we may assume that the muscle is t–cx7 in cicadas and t–cx6 in bugs. The upper ends of the muscles are in the same characteristic position in mayflies (Fig. 1.7a), stoneflies (Fig. 1.8), and other primitive insects.

The most powerful tergopleural muscle is t–p 5, 6. Matsuda suggested that this muscle arose by the splitting of a single muscle into two bundles. Such a muscle made up of two bundles is widely observed in insects but in the Neuroptera and Megaloptera there are two powerful, clearly distinct muscles. In the Orthoptera and Heteroptera only one muscle is present. Fibres of t–p5, 6 and the tergocoxal

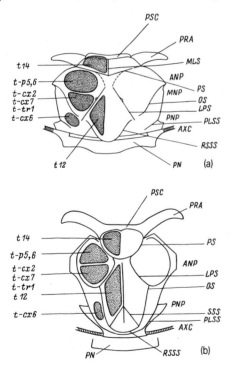

Fig. 1.8 Terga of stoneflies. (a) Mesonotum of *Yoraperla mariana*; (b) metanotum of *Kamimuria luteicauda*.

muscles extend dorsoventrally which is why this group of muscles is often called 'dorsoventral muscles'. The tergotrochanteral muscle t–tr1 is also considered to belong to this group. Apart from t–cx6, the upper ends of the dorsoventral muscles are inserted on the anterior part of the scutum. Sometimes the upper end of the tergopleural muscle and the tergocoxal and tergotrochanteral muscles unite to form a morphologically and functionally single bundle as occurs, for instance, in the Blattodea and Mecoptera; but more often only the tergocoxal and tergotrochanteral muscles are involved in this union.

The lateral parapsidal sutures are situated at the points of attachment of some of the dorsoventral muscles to the notum. In the Ephemeroptera, the lateral parapsidal suture gives an external indication of the attachment point of the powerful muscle t–p5, 6 on each side of the notum (Fig. 1.7a); in the Plecoptera, it separates the tergocoxal from the dorsal oblique muscles (see Fig. 1.8). The same arrangement of lateral parapsidal sutures has been described in some cicadas, but here the sutures mostly support the extremely powerful dorsal oblique muscles. The origin of analogous sutures in the Hymenoptera and Diptera will be discussed later.

The upper ends of a large number of small tergopleural muscles (t–p3, –4, –7, –10, and –15) occupy the most lateral position of the tergal margin. These connect the dorsal margin of the pleuron with the lateral area of the scutum and hold the thoracic box rigid. Bocharova-Messner (1968) named them the 'position muscles'.

The most severe deformation of the notum is caused by the contraction of the tergal muscles, and hence they play an important role in the formation of the tergal sutures. In the majority of winged insects, the tergal muscles are in three pairs: the dorsal longitudinal muscle (t14), the dorsal oblique (t12), and a small muscle which lies on the border between the notum and the postnotum. The attachment point of the anterior ends of the dorsal longitudinal muscle is marked by the parapsidal sutures. These sutures are present in nearly all insects but are sometimes only weakly developed

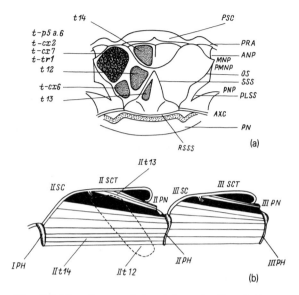

Fig. 1.9 Pterothorax of the dobsonfly *Corydalis* sp. (a) Dorsal view of the mesonotum showing the sites of attachment of the indirect flight muscles. (b) Sagittal section of the meso- and metathorax. The dorsal oblique muscle in the mesothorax is shown by a broken outline.

(as in cockroaches and orthopterans). They are absent in some Mecoptera (Hepburn 1970). In the Neuroptera, Hymenoptera, and some other insects, the sutures on both sides are medially convergent and thus bound the anteromedial part of the scutum (which is often wrongly called the prescutum). The upper ends of the dorsal oblique muscle are attached to the posterolateral parts of the notum. Their attachment points are bounded posteriorly by the branches of the scutoscutellar suture which, being medially convergent, demarcate the scutellum (see Fig. 1.5).

The small tergal muscle at the border between the notum and the postnotum is of particular interest. It is usually called muscle *t13*. Matsuda believes it to be homologous with muscle *6* in *Lepisma*, which is probably correct. However, certain facts suggest that *t13* could be the detached uppermost bundle of the dorsal longitudinal muscle *t14*:

1. The fibres of both muscles often run parallel with each other.
2. The upper end of *t13* is located medially on the notum, in contrast to the dorsal oblique muscle which lies more laterally.
3. The fibres of *t13* lie more horizontally than those of *t12*. This can be easily observed by comparing the two muscles in the illustrations of sagittal sections of the thorax of the mayfly (Fig. 1.7b) and of the dobsonfly (Fig. 1.9b).

A muscle which Brodsky (1979a) has described resembles *t13* on the border between the notum and the postnotum, in Euholognathan stoneflies. However, this muscle differs from *t13* in that its anterior end is attached to the notum on the outside of the recurrent scutoscutellar suture. An analogous muscle has been described by Matsuda (1970) in cockroaches and orthopterans. The orientation of the fibres, and the site of attachment of the anterior and posterior ends relative to the recurrent scutoscutellar suture indicate that this muscle is not homologous with *t13*. One may conclude that this muscle derives from the separation of the superior bundle of the dorsal oblique muscle. It should properly be named *t20* (Fig. 1.5) (given that the tergal muscles from *t11* to *t19* have already been described by Matsuda (1970)). Thus, while muscle *t20* is present in the thorax of Polyneoptera, in all other insects the comparable small tergal muscle is *t13*.

The involvement of the dorsal oblique and the *t13* muscles in the wing movement mechanism resulted in increased complexity of the structure of the posterior part of the notum and led to the development of the scutellum. The development of the scutellum can be divided into two stages.

1. The acquisition of the dorsal oblique muscles necessitated reinforcement of their attachment sites, which led to development of endoskeletal ridges. Now, the uniform depression of the notum which causes the wing upstroke is only possible when the upper ends of muscle *t12* are symmetrically arranged on the posterior part of the notum, as the anterior part is occupied by the dorsoventral muscles. Hence the topography of the scutoscutellar suture not only gave support

to *t12* but also affected the deformation of the notum on contraction of the dorsal longitudinal muscles resulting in development of flexible areas at the intersection of the scutoscutellar suture and the recurrent scutoscutellar suture.

2. The second stage in the formation of the scutellum was provoked by the splitting of *t13* muscle from *t14*. With the acquisition of *t13*, the scutoscutellar suture became an attachment point for the two antagonistic muscles—*t12*, the wing's levator, and *t13*, its depressor. The latter, which contracts during relaxation of *t12*, restores the dorsal curvature of the scutellum. The character of the notum deformation brought about by the contraction of the dorsal longitudinal muscles is thereby changed. When *t13* and *t14* contract, a notum with a true scutellum curves so that the median notal wing process appears to exert some influence on the first axillary sclerite. The hinges (*h1*) at the intersection of the scutoscutellar suture and the recurrent scutoscutellar suture, which are connected to the median wing processes by the oblique sutures (see Fig. 1.5), play a significant role in this tergal deformation.

The posterior scutellar margin has a panel which is laterally continuous with the axillary cord of both wings. The scutellar margin is also inflected downward, and is connected to the postnotum posteriorly by hidden suture, thus enclosing the notum in a semicircle. The ventral part of the postnotum has a two-lobed posterior phragma which extends into the body cavity. The postalar arms are made up of lateral processes of the postnotum, which are in some cases joined to the epimera.

We now come to the mode of action of the indirect wing muscles. Contraction of the dorsoventral and the dorsal oblique muscles causes an even depression of the scutum in relation to the pleural wing processes, with only minor deformation of the notum. Contraction of the dorsal longitudinal muscles which are attached to the anterior and posterior phragmas of each segment then causes severe deformation of the notum. Usually the notum arches upward between the ends of the segments, while the anterior and posterior wing processes are brought in slightly and lifted relative to the pleural wing processes. The wings move downwards.

The direct wing muscles, as has been already mentioned, are attached to separate sclerites at the wing base. The basalar sclerite provides a site of attachment for some of the pleurotrochanteral, pleurocoxal, and pleural muscles. Only in mayflies are the muscles of this sclerite weak and inadequate to cause wing movement; in all other insects the basalar muscles are well developed and function as depressors, pronators, and promotors of the wing, although the extent of their development varies considerably among insect groups. The subalar sclerite is usually operated by muscle *t-cx8*. In cockroaches and stoneflies, some of the fibres (*t-cx8a*) are inserted inferiorly on the less movable part of the coxa, at its point of articulation with the pleuron, functioning exclusively as a wing muscle. In stoneflies and members of the panorpoid complex (the Mecoptera, Diptera, Lepidoptera, and Trichoptera), muscle *t-p16* is also inserted on the subalar sclerite, while *t-p19* is inserted on this sclerite in the Hymenoptera. The significant role of the main muscle of the subalar sclerite (*t-cx8*) in flight control is clearly exemplified by the hawk moth *Manduca sexta* (Kammer 1971) and the bug *Oncopeltus fasciatus* (Govind 1972). The most powerful muscle of the same sclerite in mayflies (*t-s5*) may contract out of phase with *t14*; its role in flight control has been demonstrated using electrophysiological techniques (Brodsky 1975).

Besides the basalar and subalar sclerites, the third and first axillary sclerites are provided with muscles. Insects which fold their wings over the back in a roof-like position (with the exception of booklice) and the Diptera have both muscles *t-p13* and *t-p14* attached to the third axillary sclerite, but insects which fold their wings flat have only one of the muscles attached to it.

Although it is widely believed that the muscles of the third axillary sclerite can only act as wing flexors, there is evidence that they may exert some influence on wing motion in flight. Pfau and Nachtigall (1981) showed that muscle *t-p14* of the forewing of a migratory locust contracts tonically and controls the angle of attack of the wing. In the caddisfly *Semblis atrata*, the muscles of the third axillary sclerite contract during the wing upstroke to allow backward motion of the wing and control the rate of wing blade torsion. Moreover, in evolutionarily advanced and well-flying stoneflies, the attachment point of the upper end of *t–p16* is shifted from the subalar sclerite to the third axillary one (Brodsky 1979a).

The attachment point of the upper end of *t–p10* is transferred from the notal margin to the first axillary sclerite in the panorpoid complex. In addition, *t–p15* is attached to the posteromedian notal process by five bundles in Diptera. In mayflies, the first axillary sclerite has one slender muscle, *t–s3*, which contracts tonically and, together with another tonic muscle (*t–s4*), holds the wing steady when parachuting during the mating dance.

1.2 The axillary apparatus

The structure of the axillary apparatus is less well known than those of the skeleton and musculature. This is presumably to do with the complexity and minute size of the wing articulation and also with its location at the region between the wing and the thorax—areas of the insect body which have traditionally been studied separately. In order to understand more clearly the changes which have occurred in wing articulation in the various phylogenetic branches

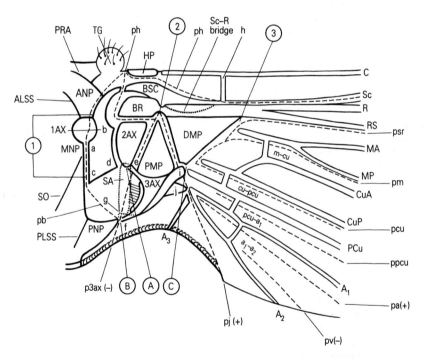

Fig. 1.10 Articulation of the wing with the tergum, and relations between the bases of the veins and the humeral plate and axillary sclerites. Here, and in the following figures, convex and concave folds (dashed lines) are indicated by the signs '+' and '−' respectively. Invisible joints are indicated by a dotted line. 1, 2, and 3 are main hinges of the wing base; A, B, and C are joints of the third axillary sclerite which allow wing folding; *a*, *b*, *c*, and *d* are four angles of the body of the first axillary sclerite; *e*, *f*, *i*, and *j* are lobes of the third axillary sclerite.

of insects, we need to be able to identify the main features of the structure and function of a generalized axillary apparatus based on the main insect groups.

1.2.1 The organization of a generalized axillary apparatus

The positions of the sclerites, folds, and points of articulation of the wing base are shown in Fig. 1.10. The tergal margin articulates with the basal structures of the wing in five places. The most anterior of these is represented by the prealar arm, made up of the lateral prescutal lobe in all insects but mantids, cockroaches, and orthopterans. The prealar arm extends downwards, and either ends loosely above the stigma or becomes fused with the episternum; in this case it forms a 'prealar bridge'. The 'tegula' is located at the point where the prealar arm deflects downwards. The most posterior site of contact between the tergum and the wing is the line which separates the notum from the postnotum, where the notum is prolonged distally in an axillary cord. The three intervening sites are on the lateral notal margin, and are named the 'anterior', 'median' and 'posterior notal wing processes'. The two anterior processes articulate with the first axillary sclerite, and the posterior process articulates with the third one. The anterior wing process is bounded by the anterolateral scutal sulcus, and the posterior by the posterolateral sulcus. The anterior wing process is separated from the median process by the tergal fissure.

Matsuda (1970) followed the basic classification of the notal margin structure of La Greca (1947), which includes five wing processes. In a number of cases (for example in stoneflies, mantids, and some other insects) Matsuda considered any roughening of the notal margin as a separate wing process. The principles of the formation of accessory processes, as well as their stages of development (which can be followed from primitive to advanced orders) suggest that the notal margin originally had three wing processes which later differentiated independently in the different groups.

The first axillary sclerite consists of a body and a characteristically curved arm. This arm always articulates with the anterior wing process, while the proximal body margin articulates, from angle *a* to angle *c*, with the median wing process (Fig. 1.10). In representatives of ancient groups such as dragonflies, stoneflies, and cockroaches, the median wing process is not differentiated, and is represented by a simple margin of varying length. The margin of the first axillary sclerite, from angle *b* to angle *d*, is intimately associated with the second axillary sclerite, while the apex of the arm articulates with the upper surface of the basisubcostale. The basiradiale joins the arm between the anterior margin of the second axillary sclerite and the basisubcostale.

The third axillary sclerite has three arms: anterior, posterior, and distal. The anterior arm (*e*) articulates with the posterior margin of the second axillary sclerite at joint A. The posterior arm is divided into superior (*f*) and inferior (*g*) lobes; the first is joined to the posterior notal wing process by joint B, while the second appears on the ventral surface of the wing base, coming to rest on the posterior margin of the subalar sclerite. The distal arm of the third axillary has two lobes, the anterior one (*i*) of which articulates with the distal medial plate and the base of the postcubital vein, forming joint C. The posterior lobe (*j*) articulates with the common origin of the anal veins. Either (or in some cases both) *t–p13* and *t–p14* are attached to the broad apodeme between the apices of lobes *e* and *f*. The triangular proximal medial plate rests on the anterior margin of the third axillary sclerite between lobes *e* and *i*. This plate articulates with the distal medial plate along the joint extending from the anterior lobe of the distal arm of the third axillary sclerite to hinge 2. The distal medial plate is frequently stretched along the basal-apical axis of the wing and connected with the medial stem base.

In the wing articulation of some insects, for example the Orthoptera, Megaloptera, Trichoptera, Hymenoptera, and others, there is an additional fourth axillary sclerite. The fourth axillary sclerite is situated between the posterior notal wing process and the third axillary sclerite. As will be shown later, the fourth axillary sclerite in the Orthoptera is a detached posterior notal wing process, while in the Oligoneoptera it is an apical piece of the lateropostnotum.

As well as the axillary sclerites, there are thickened roots of veins in the wing base. The basisubcostal sclerite, distally prolonged in a subcostal vein, is connected to the base of the radial stem by a sclerotized bridge. The costal vein meets the humeral plate. The base of the cubital stem is usually isolated on the wing membrane, whereas the base of the medial one often joins the base of the radius. The humeral vein is the most proximal of the transverse veins, and connects the costa and the subcosta on the distal side of the sclerotized bridge. The posterior end of the humeral vein is weakened by being crossed by the distal branch of the humeral fold. The outer limits of the axillary region are uncertain, but as a rule the axillary region comprises the transverse veins pcu–a_1, a_1–a_2, and also cu–pcu, the latter arising from the cubital stem where it divides into CuA and CuP. Cu always forks earlier in the hindwings than it does in the forewings. The transverse veins r–m and m–cu usually lie outside of the axillary region, but in the hindwings m–cu is frequently included in this region.

The posterior part of the flexed wing bends along folds which are located in the wing base. The proximal parts of the folds which occur in the wing surface during the wingbeat cycle are also located in the wing base. Together the folds form an intimate connection with the most flexible parts of the wing base, where the major hinges and articulations are. Vertical movement of the wing is made possible by a hinge located between the notal margin and the first axillary sclerite (the horizontal hinge, 1, in Fig. 1.10). It has one degree of freedom, acting like a door hinge. The first axillary sclerite transfers movements caused by either the elevation or depression of the notal margin to the second axillary and the basisubcostale, where wing bends along the basal fold. The vertical hinge in the radial vein base (2, in Fig. 1.10), unlike the one discussed above, allows the wing to move forward and backward. This hinge serves as the apex of the triangle which is turned upside down when the wings are folded. The concave (inferior) fold of the third axillary sclerite, the

convex (superior) jugal fold, and the neutral humeral fold all meet at this hinge. The neutral humeral fold arises from the vertical hinge and passes through the anterior wing margin between the tegula and humeral plate, projecting either between the anterior margin of the second axillary sclerite and the basiradiale or between the basiradiale and the base of the radius. The distal branch of the humeral fold arises close to the wing margin and extends between the humeral plate and the basisubcostale, crossing h close to the subcostal stem and continuing between Sc and R.

The complicated torsional hinge (3, in Fig. 1.10) lies at the apex of the distal medial plate. It provides torsion of the wing along its longitudinal axis, and is associated with deformable folds of the radial sector and of the medial and cubital. As has been already mentioned, a sclerotized bridge connects Sc and R between the vertical and the torsional hinges. The field of campaniform or thichoid sensilla which register wing torsion is frequently located near the apex of the distal medial plate.

All the sclerites of the wing base discussed so far are on the dorsal wing surface. Some of these sclerites extend on to the ventral surface of the wing. When one examines the wing articulation ventrally with the wings lifted (Fig. 1.11), three sclerites are clearly visible: the undersurfaces of the basisubcostale and of the second axillary sclerite and the inferior lobe of the posterior arm of the third axillary sclerite. The basisubcostale articulates with the basalar sclerite, which in turn is connected to the pleural wing process by a ligament. The second axillary sclerite is also connected to the pleural wing process by a ligament. The inferior lobe of the posterior arm of the third axillary sclerite rests on the caudal part of the subalar sclerite. The symmetrical arrangement of the sclerites (to which the direct flight muscles are attached) around the fulcrum of the wing presumably derives from the primitive wing apparatus. This symmetry no longer exists in the majority of recent insects: in mayflies the subalar sclerite appears to play the principal role in wing movement (Brodsky 1970b), and in Neoptera, it is the basalar sclerite, thereby causing forward migration of the pleural wing

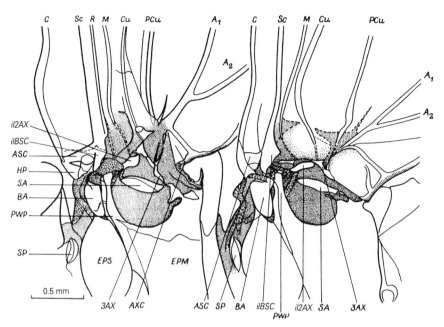

Fig. 1.11 Lateral view of the pterothorax of the alderfly *Sialis morio* with lifted fore- and hindwings, showing the system of ligaments at the wing base.

process and strong development of the inferior layer of the basisubcostale (Fig. 1.11).

We now turn to the arrangement of the tracheae in the wing base. The two branches of a Y-shaped leg trachea pass into the wing and unite to form a 'wing arc'. This is well developed in the wing base of all insects but mayflies and stoneflies. Since in mayflies the posterior branch of the Y-shaped leg trachea is degenerate, the anterior (*C*, *Sc*, *R*, and *M*) and the posterior (*Cu*, *PCu*, and *A*) stems of the veins are supplied by the anterior wing trachea. Stoneflies have presumably retained a wing base tracheation pattern which is close to the original one: the wing arc is absent, and the anterior and posterior wing tracheae enter the wing independently (Fig. 1.12). In all other insects, the wing arc remains more or less constant in structure (Chapman 1918, Comstock 1918, Whitten 1962, Brodsky 1979*b*, and others). One of the most curious variations in the tracheation of the wing base is the trio *cu*, *pcu*, and *a* (Fig. 1.12), and the oblique crossing of the distal medial plate by the

tracheae *m* and *cu* which is clearly apparent in the fore- and hindwings of the dobsonfly (Fig. 1.13)

The relative consistency in wing base tracheation in different insects may shed some light on the evolutionary relationships between some of the sclerites of the wing base and their corresponding veins. Thus the humeral plate is probably a detached proximal piece of the costal vein, the basisubcostale a detached piece of the subcostal, the basiradiale a detached piece of the radial, and the second axillary sclerite a detached piece of the medial vein. The third axillary sclerite has a complex origin; it derives from the detached bases of three veins—*Cu*, *PCu*, and *A*. The proximal medial plate is a remnant of the primitive sclerotized wing base. It became triangular when the folds which allow the anal area of the wing to be bent during wing folding developed. The distal medial plate also became triangular, because of its role in wing flexion. The first axillary sclerite is a detached notal margin which later differentiated into the body and the arm. Detachment of the notal margin and

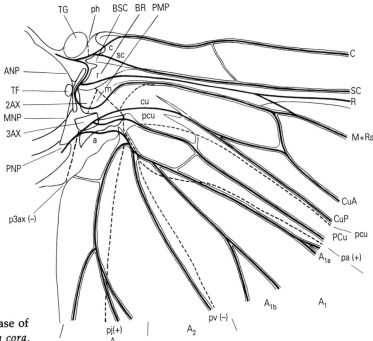

Fig. 1.12 Hindwing base of the stonefly *Sierraperla cora*, showing the main tracheal pattern of the wing base.

its transformation into the first axillary sclerite caused the elongation of the short shoulder of the wing.

1.2.2 The functioning of the axillary apparatus

Movements of the axillary apparatus during wingstrokes will be discussed in Chapter 2, which is dedicated to the work of the wing apparatus, so the following is just a summary of the work of the wing articulation. Direct and rapid transmission of the forces exerted on the wing during depression occurs via the anterior and the median notal wing processes; the basisubcostale and the second axillary sclerite appear to be important in this. The basisubcostale is in turn activated either via the basalar sclerite or, when the indirect flight muscles contract, via the first axillary sclerite. The second axillary sclerite is only activated via the first axillary sclerite. Hence, in insects such as cockroaches and orthopterans which have relatively weak

dorsal longitudinal and strong basalar muscles, the basisubcostale has become responsible for the depression and the pronation of the wing. This has shifted control of the first axillary sclerite to the anterior notal wing process. In insects in which both the dorsal longitudinal and the basalar muscles are significant for wing movement, the second axillary sclerite is most highly developed and hence the median notal wing process is too.

The muscles which produce the adduction and pronation of the wing during the downstroke are mainly those of the basalar sclerite, while those which are inserted on the subalar sclerite function as supinators and accessory depressors during the downstroke. The abduction of the wing during the upstroke is powered by elastic forces in the vertical hinge area. At the end of the upstroke, the muscles of the third axillary sclerite produce slight pronation and marked abduction, just like in wing folding. The contraction of the third axillary sclerite muscles is then followed

Look up "Ensifera" (handwritten margin note)

maybe Chapman has it? (handwritten margin note)

more thorns have inner wings: (handwritten margin note, partial)

by the contraction of the basalar muscles which complete the pronation and pull the wings forward. This explains why, if the basalar muscles did not contract at that moment, the third axillary sclerite muscles would only flex the wings.

Wing folding can be broadly divided into three main categories: flat, outlining, and roof-like. When the wings are folded flat, as in stoneflies, the wings lie flat over the abdomen, overlapping each other. The difference between flat and outlining folding, which is characteristic of the Orthoptera, is that in outlining the wing bends longitudinally, causing downward deflection of the costal margin. In roof-like folding, as in alderflies, the posterior margins of the wings meet, but do not overlap; the anterior wing margin is turned down and to the side. The transition from flat to outlining folding or vice versa is a relatively simple one, in contrast to the transition to or from either of these to roof-like folding.

The wing base sclerites move as follows during wing folding. Contraction of the muscles $t–p13$ and $t–p14$ rotates the third axillary sclerite about the axis which connects the apices of the anterior and the posterior arms ($e–f$, see Fig. 1.10, p. 16) and elevates its distal arm, while the triangle formed by the hinges A–2–C rotates downwards. Elevation of the distal arm produces a deep convex fold between the two medial plates from hinge 2 to hinge C. This movement is accompanied by abduction and pronation of the wing, while the clavus becomes flattened and the postcubital vein base depressed.

1.3 The wings

Wing structure has been studied for insect taxonomy, especially for fossils, ever since Linnaeus (1707–78). As well as the extensive descriptive literature, a few summaries of wing structure have been published: Comstock (1918), Wootton (1979), Brodsky (1987), and others.

1.3.1 Wing shape

In insects which flap their fore- and hindwings independently in flight, the most usual wing shape is a stretched ellipse with a straight anterior margin (Fig. 1.14, a–c). The wing is widest at its distal half. In such wings symmetry is common, with the line which connects the base of the wing with its apex dividing the wing into two equal parts. This arrangement is characteristic of primitive moths, Mesozoic scorpionflies, and others. The shape of the wings of primitive caddisflies, recent scorpionflies (Fig. 1.14d), and dipteran moth flies is also nearly symmetrical. In insects in which the hindwings are the principal flight organs (orthopterans, cockroaches, earwigs, and mantids) and also in cicadas, bugs (Fig. 1.14e), and beetles, the hindwing broadens at its anal part and thus becomes a right-angled or even acute-angled triangle, its anterior margin forming the hypotenuse.

In insects with coupled wings, the shape of the wing pair on each side also resembles a right-angled triangle. In general, insects with short oval abdomens have relatively broad wings, whereas those with slender and elongated abdomens have relatively narrow wings. The most significant deviations from these two basic elliptical or triangular types occur in those insects which use both active flapping flight and passive flight (gliding and soaring).

1.3.2 Venation

Let us begin with a generalized wing, including all the longitudinal veins and the main folds (Fig. 1.15). As the wings of some insect groups fold in a fan-like manner, with the longitudinal veins running along the crests and troughs, veins on crests are referred to as convex and those in troughs are referred to as concave. The marginal or submarginal *costa* (C) is a simple convex vein, with a bundle of short branches at its base in some Palaeozoic insects and Orthoptera. Branches off the costa are called *precostal* veins. These precostal veins are more strongly developed—longer and more numerous—the wider the precostal area. In Ensifera (Orthoptera), the position of the forewings at rest is correlated

prim. (handwritten)  *adv.* (handwritten) *vs* (handwritten)

he says (handwritten)

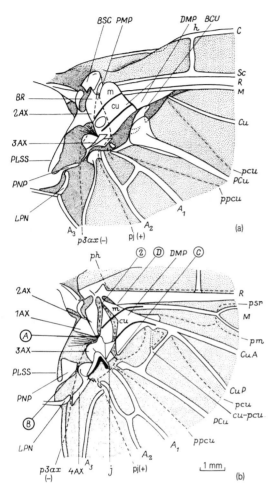

Fig. 1.13 Bases of (a) the forewing and (b) the hindwing of the dobsonfly *Corydalis* sp., showing the oblique crossing of *DMP* by the *m* and *cu* tracheae. (For explanation of labels A–C and 2, see Fig. 1.10; for description of hinge D, see Fig. 9.16).

margin and frequently branches at the tip. The *radial* vein (*R*) forks into an undivided convex *radius* and a concave *radial sector* (*Rs*). Branches off the radial sector emerge on the wing tip in parallel. Typically, the *media* (*M*) forks into two equal branches, the *anterior* (*MA*) and the *posterior* (*MP*) *medias*. The latter is sometimes bounded posteriorly by the medial fold and is therefore concave. A strong transverse vein (*m–cu*) arises from the media near its branching point

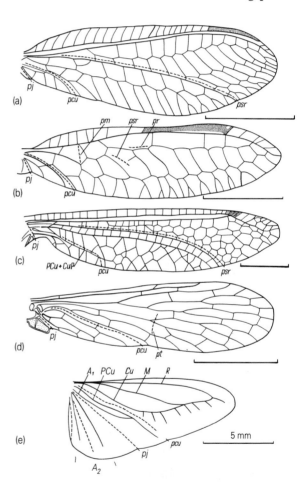

Fig. 1.14 Examples of common insect wing shapes. (a) Forewing of the green lacewing *Chrysopa perla*; (b) forewing of mantispid *Mantispa styriaca*; (c) forewing of *Ululodes senex*; (d) forewing of the scorpionfly *Panorpa communis*; (e) hindwing of the waterscorpion *Nepa cinerea*.

with the degree of development of the precostal veins. Thus, in those primitive Ensifera which keep their folded forewings slightly depressed (see Fig. 7.15a, p.129), the broad precostal areas enclose the sides of the thorax and have numerous precostal veins.

The next longitudinal vein, the *subcostal* (*Sc*), is concave and lies parallel to the anterior wing

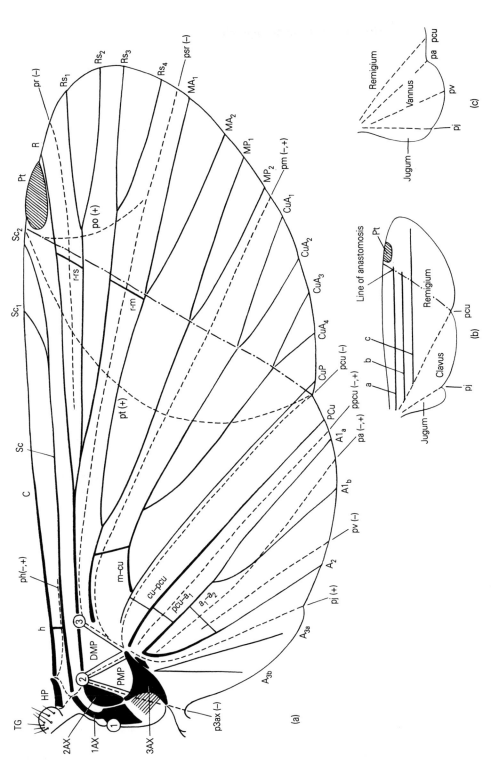

Fig. 1.15 The principal features of a generalized insect wing. (a) General view of the wing showing the axillary sclerites, main longitudinal veins, and folds. The anastomosis line is shown by a dot-dash line. *Pt* is the pterostigma. (b) Regions of the wing showing the principal axes: *a*, wing rotational (torsional) axis; *b*, wing mass axis; *c*, axis of aerodynamic pressure. (c) Expanded anal lobe of the wing showing the arrangement of the folds used in wing folding.

into *MA* and *MP*. *m–cu* runs obliquely and posteriorly, and is crossed by the medial fold. This vein is characteristic of the wings of the majority of insects and, as will be shown later, evolved very early in the phylogenetic history of winged insects. The Moscow palaeontologists (see Rasnitsyn 1980*b*) have suggested that *m–cu* constitutes the base of the most proximal branch of the medial vein (M_5), and is fused with *CuA*. If this view is correct, the media appears to have three main branches: M_5, *MA*, and *MP*. In orthopterans Sharov (1968) designated M_5 as *MP*.

The next vein, the *cubitus* (*Cu*) like the media forks into the *anterior* (*CuA*) and the *posterior* (*CuP*) *cubiti*: the anterior branch is usually convex, whereas the posterior one is concave and does not fork. The cubital vein is followed by the *postcubitus* (*PCu*) which is the furthest anterior and the strongest vein of the clavus. *PCu* is thinner in the hindwings and becomes barely distinguishable from the *anal* veins (*A*) located behind it. At its base, the postcubitus is associated with the anterior lobe of the distal arm of the third axillary sclerite and is supplied with an independent trachea which originates in the cubital (Fig. 1.12, p.20).

The question of the number and homologies of the anal veins is particularly problematic. The presence of the jugal veins at the place where the wings are folded enabled Martynov (1924) to separate the Neoptera from the Palaeoptera. In spite of mistakes, such as having miscounted the number of jugal veins in Orthoptera, Martynov's ideas appear to have been productive, and the problem the of jugal veins is closely connected with that of the number and homology of the anal veins. The anal veins—those located on the wing membrane between *PCu* and the jugal fold—are articulated by a single joint with the posterior lobe of the distal arm of the third axillary sclerite. The base of the vein (or veins) posterior to the jugal fold is either unattached (Fig. 1.15a), or else the vein itself originates at the base of the anal veins (Fig. 1.10, p.16). The veins of both types are supplied with a separate tracheal branch which divides into two—distal and proximal—at the common base of these veins (Fig. 1.12, see p.20). Thus one anal vein clearly divided into two main branches, the anal

and the jugal ones; the anal vein later subdivided into two anal veins, A_1 and A_2.

The wings of the majority of recent insects have two anal veins. Stoneflies have only one, three-branched anal vein in the forewing, which is located distally from the jugal fold (Fig. 1.16). These three branches correspond exactly with the two anal veins of the hindwing, A_1 and A_2, which are separated by a concave (vannal) fold. In the forewing of crickets the anal vein has the same structure as in stoneflies (Fig. 1.17c), but in the Caelifera (Orthoptera), there are two distinct veins on the clavus, the anterior one of which forks extensively posteriorly, whereas the posterior vein is unforked. On the clavus of cockroaches there are numerous parallel veins, divided into two bundles; the posterior bundle resembles the structure of the anal fan

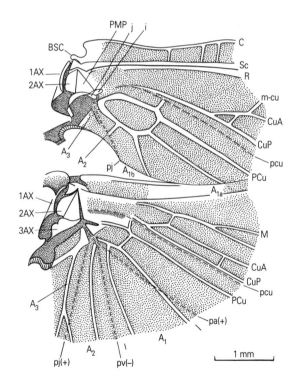

Fig. 1.16 Bases of the fore- and hindwings of the stonefly *Yoraperla mariana*. The anal stem has the same branches in both the forewing and the hindwing, namely A_1, A_2, and A_3.

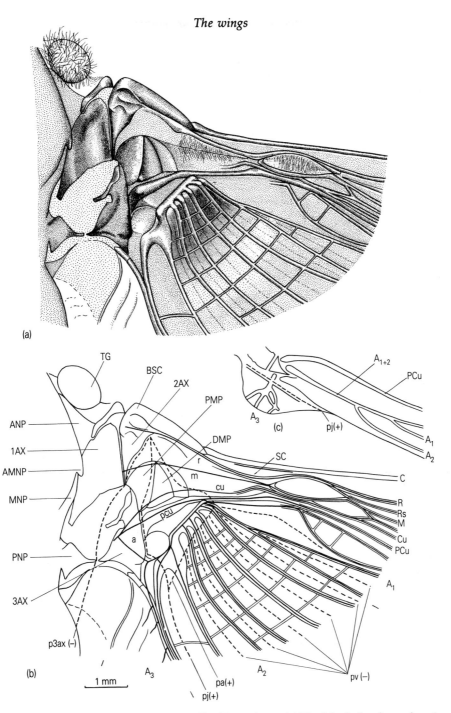

Fig. 1.17 Base of the hindwing of the cricket *Gryllus bimaculatus*. (a) The hindwing base showing its general structure. (b) The hindwing base showing the arrangement of sclerites, vein bases, folds, and tracheae. (c) clavus of the forewing (not to scale). The branches of the anal stem are the same in the fore- and hindwings, namely A_1, A_2, and A_3.

of the hindwing. In the Oligoneoptera there are two anal veins on each of the fore- and hindwings, just above the jugal fold, whereas in the Paraneoptera, they occur in this region, but only on the forewings. Hence all Neoptera have two anal veins on the clavus. On the hindwings the anal vein A_1 is forked in most species, and is separated from A_2 by a concave fold (Figs 1.16 and 1.17). A_2 is also highly subdivided, forming a regular comb on the hindwings of the Blattodea and the Orthoptera (Figs 1.17 and 1.18).

The vein (or veins) which is situated on the wing fold itself is indirectly connected to the third axillary sclerite, and is supplied by a proximal twig of the anal tracheal branch (Figs 1.12 and 1.17). This vein is presumably the most extreme posterior branch of the anal stem, and therefore constitutes 'A_3'. A_3 is clearly present in most of the Neoptera (Fig. 1.13, p.22), but is not conspicuous in the Palaeoptera. It is dificult to homologize all of the anal veins in the Palaeodictyoptera as their wing bases are crossed by a jugal fold which participates in the deformation of the wing during supination. In fossil mayflies and dragonflies either folds or thickening are visible, and it is not possible to establish whether these are branches of A_3.

The main function of the veins is to provide mechanical stiffening for the wing. The loadings imposed upon the wing during flight may be substantial, as a result of aerodynamic pressure at the middle of the up- and downstrokes and inertial forces at the stroke reversal points. In a wing flapping with a sinusoidal motion, the torques which cause angular acceleration are changing constantly. The centre of mass of an insect's wing is often behind its rotational (torsional) axis (Norberg 1972b). If such a wing accelerates the inertia of the wing will generate torque around the rotational axis, which should peak at stroke reversal when acceleration is high. The pronational–supinational (basal–apical) axis of rotation is shifted forward (Fig. 1.15b) and aligns approximately with the radial vein; this vein is also the main mechanical axis of the wing, through which the forces produced by the wing muscles are transmitted to the wing blade. The line along which the wing's centre of mass lies passes between the rotational axis

and the midline of the wing (Fig. 1.15b), where thick and heavy veins are clustered. Similarly, the centre of pressure of aerodynamic forces also tends to act behind the rotational axis of the wing. Aerodynamic forces will therefore tend to produce torque, which should rise to a maximum in the middle of each wingbeat, when the wing velocity is greatest. The location of the line along which the centre of aerodynamic pressure lies on the wing depends on the angle of attack of the wing. In relatively primitive insects, this line lies approximately along the midline of the wing (Fig. 1.15b), while in advanced forms, it is shifted forward to the leading edge of the wing.

As has already been mentioned, the costal vein has the most anterior position on the wing, followed by the subcostal vein, which is often fused to the costal vein at the wing tip. In accordance with its mechanical role, the radius runs parallel to the leading wing edge. Other veins radiate from the base to the apex and posterior margin of the wing (Fig. 1.15a), with a tendency to fork more towards the posterior margin of the wing. The vein branches spread out on the posterior and terminal margins of the wing in equidistant parallel rows. The branches become thinner at the posterior margin, thus increasing wing flexibility from base to tip and from the anterior to the posterior margin of the wing. The transverse veins and curves of the main longitudinal veins form concentric rows parallel to the wing margin (Fig. 1.14a,b). Thus the venation forms a close network which provides a framework for the wing combining rigidity and flexibility. This structure alone would be sufficient to generate aerodynamic forces under steady flight conditions. However, inertial forces arising out of the wingstroke make extra demands on insect wings.

1.3.3 The system of folds

Functionally the wing folds may be classified into three separate groups which provide:

(1) deformation of the flapping surface;
(2) bending of the posterior part of the wing at folding;
(3) compact packing of the expanded anal area of the hindwing (the vannus).

The development of folds in the first group was necessitated by the considerable inertial forces which act on the wing at the stroke reversal points. The folds may lie along the wing veins or across them, morphologically dividing up the various areas of the wing (Fig. 1.15b). The *cubital fold (pcu)* is the most important longitudinal fold. It starts at the wing root, and runs from the anterior lobe of the distal arm of the third axillary sclerite to the posterior margin of the wing between *CuP* and *PCu*, separating off the posterobasal triangular lobe of the wing (Fig. 1.15b). This results in narrowing of the wing base, facilitating its rotation along the longitudinal axis during pronation and supination. Thus, *pcu* divides the wing plane into two unequal parts, the remigium and the clavus, which thereby derive some autonomy of movement. The separation of the clavus is associated with the development of the postcubitus into its main supporting vein, and it is therefore unforked. On the posterior margin, the clavus is separated from the remigium by a small indentation.

The cubital fold evolved in insect wings very early. That *pcu* was present in ancient Carboniferous insects is evident from the well pronounced clavus, the indentation of the posterior margin, and the characteristic network of intercalary veins between *CuP* and *PCu* (see Fig. 1.19a). The cubital fold retains its typical position in the forewings of nearly all recent insects. In some cases it has been transposed, in association with various modifications of the wings, but in the wing base it always originates between the bases of *Cu* and *PCu*. Sometimes it runs right along *CuP*, rendering it in an inferior (concave) position (as in mayflies, primitive stoneflies, some Neuroptera, and others).

The homologue of the clavus in the hindwings is the anal lobe, which, when expanded, is called the anal fan or the vannus (see Fig. 1.15b,c). Expansion of the anal lobe of hindwings was necessitated by the requirements of aerodynamics; the vannus is more flexible than the clavus, and its mechanics differ markedly from that of the clavus. It inevitably influences the character of *pcu* and *PCu*. *PCu* loses its main supporting role at the rear part of the wing, becomes thinner,

and sometimes partly reduced, and *pcu* ceases to act as the principal dividing line between two wing areas with different flight functions. In the expanded cubital field of cockroach hindwings, *pcu* encroaches, but still lies along *CuP* (Fig. 1.18). In the forewings of these insects, *CuP* is strongly concave and forms the anterior margin of the clavus (an arc-shaped suture).

The next important longitudinal fold which originates from the distal arm of the third axillary sclerite is the *radial sector fold (psr)*. This cuts across the bases of the medial and the cubital veins which become thinner, having lost their connection with their sclerites. The medial and the cubital veins are therefore left hanging; and thereby acquire some autonomy of movement. Like *pcu*, *psr* is always concave—that is, during longitudinal bending of the wing, the areas divided by *psr* form an angle whose apex is directed downward. The fold may be branched.

The radial sector fold is primarily characteristic of broad hindwings (Fig. 1.19b) and is less widespread than *pcu*. Besides these two principal longitudinal folds, some other folds occur on insect wings from time to time: the *radial (pr)*, the *medial (pm)*, the *anterocubital (pcua)*, and the *postcubital (ppcu)* (Fig. 1.15a). The medial fold is the most interesting one. Although like other longitudinal folds it is morphologically inferior (i.e. concave), in alderflies and some caddisflies the fold which forms along *pm* when the wing is bent points upward. In hover flies and broad-winged butterflies the medial fold allows both upward and downward bending. However, in the Neuroptera, the medial fold always points downwards, which is why it is considered to be *psr* (see Fig. 1.14a, c, p.22).

Transverse bending of the wing often occurs along a fixed line corresponding to the *transverse fold (pt*, or nodal line) which crosses the wing plane diagonally or semicircularly (Fig. 1.15a). The transverse fold is always superior (i.e. convex) and is present in a variety of insects. The *oblique fold (po)* is also superior (convex) and occurs less commonly, usually in the Hymenoptera.

Folds of the second functional group allow the posterior part of the wing, which is folded over, to bend. There are two folds in this category:

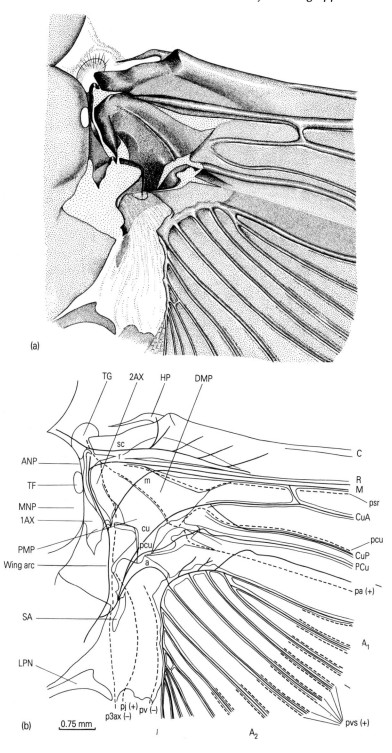

(a)

(b)   0.75 mm

TG   2AX   HP   DMP

sc

r

ANP

TF

MNP

1AX

PMP

Wing arc

SA

LPN

m

cu

pcu

a

C

R
M

psr

CuA

pcu

CuP
PCu

pa (+)

A₁

pj (+) pv (−)
p3ax (−)

pvs (+)

A₂

Fig. 1.18 Hindwing base of the cockroach *Periplaneta americana*. (a) The hindwing base showing its general structure. (b) The arrangement of sclerites, vein bases, folds, and tracheae. *pvs* is a supplementary vannal fold uniquely characteristic of cockroaches and mantids.

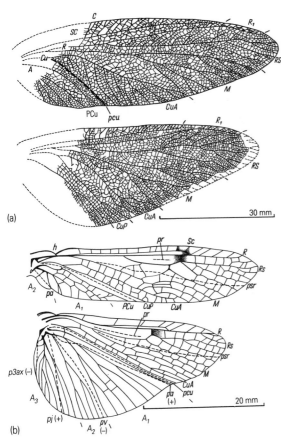

(a)

(b)

Fig. 1.19 Wings of different insects. (a) The extinct protopteran *Zdenekia grandis* (after Kukalová 1958). (b) The recent stonefly *Allonarcys sachalina*.

The folds of the anal area of the wings have been given different names by different authors (Snodgrass 1935, Wootton 1979, Brodsky 1987, 1988, and others). The line along which the jugum folds should be called the 'jugal', as it is homologous in all wing-folding insects, although it is partly reduced in the hindwings of the Orthoptera, and is almost eliminated in those of the Blattodea (Fig. 1.18). In both these groups and in stoneflies the vannus bends along the anal fold (Figs 1.16–18). Supplementary folds in the hindwings of some caddisflies and moths are situated distally, and the jugal is still the major fold. In these insects the postcubital, the anal, or a new fold in the cubital field may act as supplementary folds.

The structures and positions of the folds of the third functional group are more various than those of the second one, demonstrating their secondary origin. There is only one concave *vannal fold* in the hindwings of stoneflies (Fig. 1.19b). This fold divides into a number of folds, drawing together the veins of the anal fan. The vannal folds of Orthoptera presumably originated this way (Fig. 1.17). There are a number of supplementary folds (*pvs*) in cockroaches and mantids (Fig. 1.18). These are slightly convex and lie on both sides of the veins of the fan, which are separated by a membranous area, and form a flap when the vannus folds. On the expanded bases of the hindwings of some bugs, beetles, and cicadas, concave vannal folds alternate with convex folds, thus enabling the anal part to pack compactly inside the folded lobe. Convex folds are always more marked than concave ones.

The system of folds changes the mechanical properties of the wing, giving greater flexibility to some areas of the plane relative to others. The most flexible part of the anterior margin is the area adjacent to the tip of the subcosta, from which the transverse fold usually originates. If there is a pterostigma on the wing, it always lies more distal than the anterior end of the transverse fold. The place where the posterior margin merges with the cubital fold is the most flexible part of the wing. An imaginary line connecting the most flexible part of the anterior and the posterior margins is named the 'anastomosis' (Fig. 1.15b). The position of this

the convex *jugal fold* (*pj*), and the concave *third axillary sclerite fold* (*p3ax*). These two form a functional pair, which act as the sides of a hinged triangle, the base of which is formed by the third axillary sclerite (Fig. 1.15a). The concave fold occurs in all insects capable of folding their wings. The convex fold in the forewings originates from the vertical hinge, projects between the medial plates and then, having rounded the distal arm of the third axillary sclerite, terminates at the wing edge, characteristically marked by a small cavity. On the clavus, *pj* divides off a triangular lobe, the jugum (Fig. 1.15b), which is effectively the folding section of the clavus.

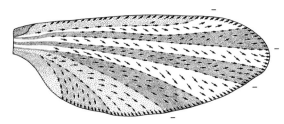

Fig. 1.20 Orientation of the ridges and grooves (−) on the wing surface. Short arrows indicate the orientation of microstructures such as hairs, scales, spines, etc. (after Bocharova-Messner 1982).

line on the wing relative to the anterior margin (perpendicular, and slightly sloping) serves as an indicator of the mechanical properties of the wing which characterize its deformation during strokes. The most flexible part of the anterior margin is always morphologically distinct. Thus, in insects with a long subcosta, approximately at the point where *Sc* ends in stoneflies, there is a coloured flexible area which is often erroneously regarded as the pterostigma. This phenomenon is common in many insects with a long subcosta: Neuroptera, Ephemeroptera, Permothemistida, and others.

1.3.4 Relief

A complex relief is characteristic of the surface of the insect wing; comprising a number of concave and convex veins, grooves, and corrugation of the wing membrane (the 'macrorelief'), as well as various articulated and unarticulated outgrowths

from 1–3 to 120 μm high, in the form of spicules, spines, hairs, and scales (the 'microrelief'). The main components of the macrorelief are the alternating concave and convex veins or 'corrugation'. This corrugation is most highly developed at the base of the anterior part of the wing: *C* is convex, *Sc* is concave, *R* is convex, while other veins are comparatively neutrally positioned. Sometimes *CuA* may be convex, and *CuP* concave. In the majority of relatively primitive insects the corrugation is highly developed. It is extremely well defined at the wing base and gradually diminishes towards the tip. Such corrugation of the wing surface results in the development of various grooves which project from the base to the tip of the wing, diverging slightly forward and much more strongly backward (Fig. 1.20).

The microrelief is best developed in representatives of historically young groups, and comprises outgrowths of various shapes and sizes. Their orientation appears to be related to the macrorelief (Fig. 1.20). Three major features characterize the distribution of the outgrowths on the wing surface:

1. The microstructures form rows of longitudinal orientation in the medial region.
2. In the marginal zone (the terminal and posterior margins), these outgrowths constitute microgrooves which run perpendicular to the margin.
3. In the costal field the outgrowths are oriented obliquely forward, and distally, at an angle of approximately 50–60° to the wing margin.

2

Mode of action of the wing apparatus

When the insect is flying forward, without accelerating, a balance of forces is observed, such that lift counterbalances body weight and thrust compensates total drag (Fig. 2.1). The long axis of the body is at an angle to the direction of flight, and the body drag experienced by the flying insect is a function of both body shape and conditions of airflow. A study of airflow around different insects by Brodsky and Ivanov (1974) showed that vortices are formed behind any body shape. The most streamlined shape was found to be the hawk moth *Amorpha populi*, but it is still far from ideal as the body thickness is greatest in the front half which causes the

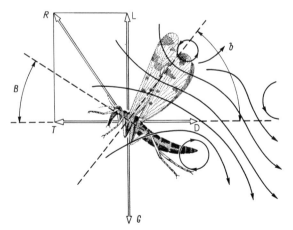

Fig. 2.1 Forces acting on a flying scorpionfly. The longitudinal body axis forms an angle *B* with the horizontal. *b* is the stroke plane angle relative to the horizontal. The beating wings create lift and thrust as the air is thrown backward and downward, as indicated by the thin arrows.

airflow to stall at a speed of 3 m s^{-1}.

The legs of insects are folded during flight, and held against the body (Fig. 2.2); sometimes they may be stretched slightly sideways. In most insects, the hindlegs are drawn up along the body. Only in longhorned grasshoppers, locusts, and mantids are they folded. The hind and sometimes the midlegs may diverge to the side of the turn when the insect changes direction. The antennae of the flying insect are stretched obliquely forward, symmetrical to the oncoming air current. The long antennae of some caddisflies and moths are stretched out sideways in flight, and their tips may be bent backward.

The long abdomens of insects with low wing-beat frequencies may vibrate in the vertical plane during flight, in antiphase with the wing strokes (Fig. 2.3). In ichneumon wasps and mayflies the abdomen is frequently elevated and held forward; with alteration of the flight direction the abdomen bends towards the turn. By moving the abdomen up or down these insects can change the body drag substantially, as observed during the flight of the mayfly *Ephemera vulgata* (Brodsky 1971). Such changes in abdominal position are accompanied by a shift in the centre of gravity forwards or backwards which causes an alteration in body angle. In such insects, if the stroke plane angle does not vary relative to the longitudinal body axis, the direction of the full aerodynamic force generated by the wings changes its orientation relative to the horizontal. Insects with short abdomens, such as flies, use the hindlegs for this purpose (Vogel 1967*a*). The above is a complete account of how the abdomen and hindlegs are used for steering during flight—most changes in flight trajectory are accomplished by the wings.

2.1. The role of the axillary apparatus during flight

The wings are moved about the body along a particular trajectory, in which their movements consist of back-and-forth and up-and-down oscillations, the rotation of the wings around the longitudinal axis (pronation and supination), and deformation. Movement is transmitted from the muscles and the thoracic skeleton to the wings via their articulation with the thorax. The events which occur in the wing base during flight have been most extensively studied by Brodsky and Ivanov (1986) in the large caddisfly *Semblis atrata* from the primitive family Phryganeidae. The base of the forewing of this species is shown in Fig. 2.4(a).

Each wing stroke can be divided into four distinct stages:

(1) depression and turning forward;
(2) turning backward and beginning supination;
(3) elevation and end of supination;
(4) pronation.

Let us examine the processes which take place at each stage of the stroke.

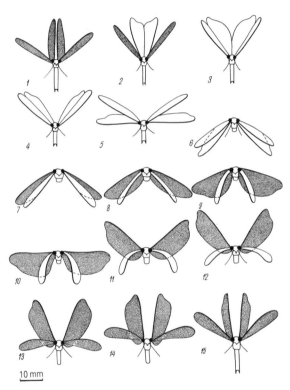

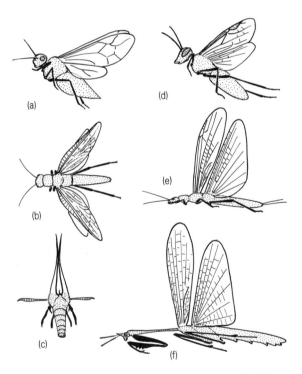

Fig. 2.2 Characteristic flight postures of different insects. (a) Psocid *Amphigerontia intermedia*; (b) alderfly *Sialis morio*, plan view; (c) skipper *Thymelicus lineola*, rear view; (d) sawfly *Rhogogaster viridis*; (e) stonefly *Amphinemura borealis*; (f) mantid *Empusa pennicornis*.

Fig. 2.3 Flight of the mantid *Fischeria baetica* (front view); consecutive film tracings (1–15) from a single wingbeat. The wingbeat frequency (n) is 25 Hz, the stroke angles (A) are 70° and 110° for fore- and hindwings respectively, the stroke plane angles (b) are 50° and 70° for the fore- and hindwings respectively, and the body angle (B) is 20°. The undersurface of the wings is shaded.

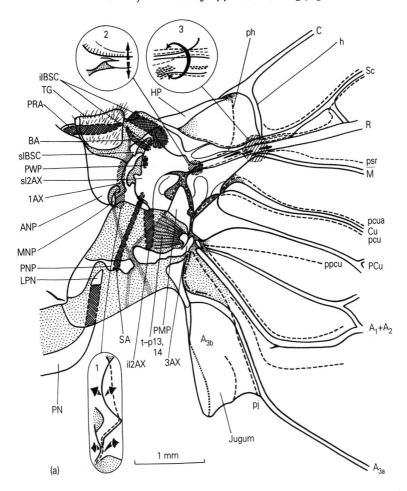

Fig. 2.4 Kinematics of the forewing joint during the stroke cycle of the caddisfly *Semblis atrata*. Both the hinge areas and structures beneath the wing plane are shaded. Arrows in (b), (c), (d), and (e) indicate the direction of displacement of the sclerites. (a) Dorsal view of the forewing base showing the arrangement of sclerites, hinges, veins, and folds on the extended wing. The arrangement of sclerites in the wing base is shown in (b) at the stage of wing depression and turning forward, in (c) when the wing turns backward and begins to supinate, in (d) at wing elevation and the end of supination, and in (e) at wing pronation.

The stroke cycle begins with wing depression and turning forward (Fig. 2.4b). The dorsal longitudinal musculature moves the anterior phragma backward and the postnotum forward. At the same time the notum arches upward, most pronouncedly at the border between the scutum and the scutellum and also in the region of the median notal process. The horizontal hinge between the notal margin and the first axillary

sclerite then begins moving: the first axillary, which was turned upward distally, rotates about its articulation, becomes horizontal, and then bends downward. While the first axillary sclerite rotates, the median notal process lifts off the notum, much as a book opens. Since the first axillary is closely associated with the second axillary sclerite (forming a continuous plate), the movement of the first axillary is transmitted to

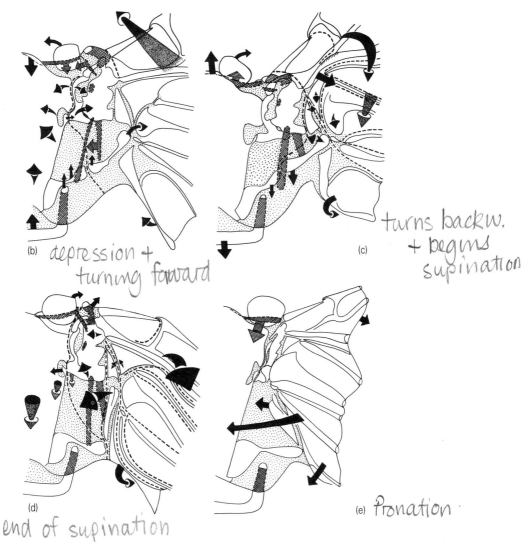

(b) depression + turning forward

(c) turns backw. + begins supination

(d) end of supination

(e) Pronation

the second, thus causing a strong deflection of its distal margin. As the wing goes down, the ligament which connects the arm of the first axillary sclerite with the anterior notal process becomes stretched.

In parallel with the process of depression, the wing rotates forward about the vertical hinge and its costal margin bends downward. The latter movement is assisted by the torsional hinge. Moving forward, the lateropostnotum exerts pressure on the posterior notal wing process, turning it forward. As a result of this, the base of the third axillary sclerite travels forward and its distal arm is depressed at the same time as the wing plane. The prealar arm then moves backward and the tegula approaches the notum. This restricts the membranous areas between the tegula and the humeral plate and also between the postnotum and the scutellum, causing depression of the subalar sclerite, so that the basalare sclerite moves inward towards the body.

At the second stage of the stroke cycle (Fig. 2.4c), in the vertical hinge there occurs a rapid reverse of the wing; the tegula moves closer to the wing, the prealar arm and the anterior phragma are restored to their original positions, and the postnotum moves backward. The jugum bends along the jugal fold and a strong convex fold is produced along the line of articulation of the medial plates. The posterior notal wing process revolves backwards, coinciding with the elevation of the distal arm of the third axillary sclerite and the bases of the veins which are connected with it. The muscles of the third axillary sclerite contract at the very beginning of supination and, judging from frame-by-frame film analysis, remain contracted throughout the wing elevation stage.

The anterior margin of the wing turns upward during supination but as the articulation area remains flat, the wing is twisted at its base. This produces a deep depression along the anterocubital fold, and corresponding depression in front of the cubital vein. To a lesser extent the wing is also twisted by the basal parts of the radial sector fold and by the cubital fold. The radial vein twists near the torsional hinge, while the humeral fold provides elevation of the distal part of the costal field of the wing. The abduction of the wing and its supination are two stages of a single process which merge smoothly into one another, and continue with the elevation of the wing.

In elevation (Fig. 2.4d), the supine wing elevates and moves slightly backward. The dorsoventral muscles flatten the notum and depress its margin. At the same time, the distal margin of the first axillary sclerite turns upward, in turn moving the second, superior layer of the basisubcostal sclerite, and hence the wing turns upward too.

The distal arm of the third axillary sclerite first turns up and then in; using cinefilm, the muscles of the third axillary sclerite can be clearly seen swelling up as they contract. The tegula then returns to its original position, and the median wing process approaches the notum. The subalar sclerite moves upward, while the basalare sclerite moves away from the body.

In the final stage (Fig. 2.4e), the wing moves slightly backward, its anterior margin turns downward, and the posterior margins of both forewings shift toward each other. Presumably this movement is provided by the continuing contraction of the muscles of the third axillary sclerite. In addition, the subalar sclerite may contribute to the backward movement of the wing. The basalar sclerite and prealar arm may also take part in the turning downward (pronation) of the anterior margin of the wing.

Read above 5 times ... still can't follow it.

2.2 Wing kinematics

The movements of the sclerites of the wing base described above result in the wing stroke. The French physiologist Marey (1869) was the first person to follow the wing movement of a flying insect, and discovered the figure-of-eight curve described by the wing tip, which is well known to all entomologists and has taken root in numerous textbooks and manuals. Since then, students of insect flight have made a detailed study of insect wing kinematics, sometimes using very sophisticated devices. The characterization of wing kinematics necessitates the involvement of a number of stroke parameters, described below.

2.2.1 Stroke parameters

Wingbeat frequency comes closest to characterizing insect flight. It varies a great deal between species, and is sometimes very high. A detailed study of wingbeat frequency was carried out by Sotavalta (1947). Using mainly acoustic methods, he calculated the wingbeat frequency for a large number of species—representatives of twelve orders. The oscillations of the wings in flight are complex, and the pitch of the sound depends on both wing oscillation frequency and wing harmonics, as well as on other sound sources such as the flexible parts of the wing, the halteres of flies, etc. This might explain why

Once heard a story that Nagtigall had perfect pitch and could tell freq. by listening to insect

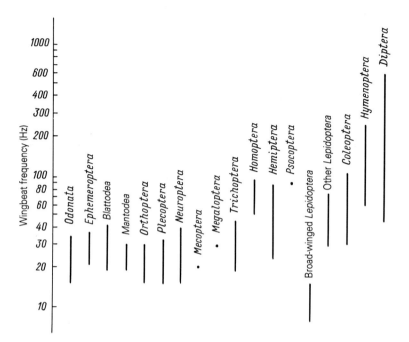

Fig. 2.5 Wingbeat frequencies in insects of different orders; confirmation of Sotavalta (1947).

in some cases Sotavalta's wingbeat frequencies appear to have been overestimated (e.g. those for mayflies, dragonflies, flies, mosquitoes, and others). Comparison of the characteristic wingbeat frequencies of insects in the main orders (Fig. 2.5, data for very small insects are not included) shows clearly that the winged insects can be classified into two groups: slow and fast flapping insects. The slow flappers' lowest wingbeat frequency is 17 Hz (with an average flight wingbeat frequency of 30 Hz), while the fast flappers' lowest wingbeat frequency is 30 Hz and more. The influence of body size is marked within a group of insects with the same wing form; relatively small insects have higher wingbeat frequencies. Thus, small cockroaches from the genus *Ectobius* are characterized by a wingbeat frequency of nearly 40 Hz, whereas giant bugs from the genus *Lethocerus* have a wingbeat frequency of only 25 Hz. In addition, wingbeat

frequency differs slightly between sexes and varies with surrounding air temperature. However, in the desert locust, individual variation in wingbeat frequency varies only 20–30% from its average value, and this variation is even less in swift-flapping insects.

When studying the flight of the blow fly *Calliphora*, Nachtigall and Roth (1983) discovered a correlation between lift values and wingbeat frequency. The same correlation was found for the house fly (Spüler and Heide 1978) and the small tortoiseshell (Gewecke and Niehaus 1981). In the water beetle *Dytiscus marginalis*, wingbeat frequency increases linearly with increasing air speed above 2.5 ms^{-1} (Bauer and Gewecke 1985). Nevertheless, this does not mean that wingbeat frequency is widely used by insects to vary the force produced. On the contrary, flying cockroaches and locusts tend to maintain a stable rhythm of oscillations. Still, even a relatively small change in wingbeat frequency can cause a significant increase in lift in fast-flapping insects (Nachtigall and Roth 1983). Wingbeat frequency in Diptera is controlled by the muscles which regulate the tension of the thoracic walls (Nachtigall and Wilson 1967)—one might infer that insects with high wingbeat frequencies actively use this as a means to control the forces produced.

Stroke angle (wingstroke amplitude) is defined as the angle subtended by the most extreme positions of the wing in the stroke plane. Stroke angle varies a great deal between species, but is on average about 120–130°. The largest value of this parameter occurs in insects which clap their wings together at the extreme upper and lower positions (some broad-winged butterflies, and most beetles). Wide stroke angles also occur in the mayfly *Ephemera vulgata* (160°), the house cricket (175°), and some stoneflies and lacewings (150°)—i.e. in relatively primitive insects. By contrast, all anisopterous Odonata have an extremely small stroke angle. It is also small in all but the most minute Diptera. Mosquitoes have by far the smallest stroke angle: 25° in males and 45° in females (Sotavalta 1952). Though stroke angle in these free-flying insects was estimated visually, the data obtained presumably approximate well to reality.

As a rule, the stroke angle declines on the inside of a turn and increases on the outside, but in the bug *Anoplecnimes curvipes* only the inside-turn decrease occurs. In the mantid *Empusa pennicornis*, as flight speed increases (as the legs gradually take up their flight position), the stroke angle of the forewings decreases (from 80 to 40°), whereas the stroke angle of the hindwings remains constant (at 130°). When green lacewings switch from level forward flight to rising flight, their stroke angle increases considerably. Field observations show that search and migratory flights of broad-winged butterflies are characterized by longer strokes relative to those of feeding flights. However, flying locusts always keep their stroke angles constant. This is because in the Orthoptera, control over the forces generated is achieved not just by alteration of the stroke angle, but also by varying the angle of attack of the forewings (Svidersky

1973). This distinguishes the Orthoptera from all other insects that correct their flight by altering the stroke angle.

Stroke plane angle is currently defined as the angle between the stroke plane and the vertical axis, on which the centre of gravity of the body lies. It is also informative to consider the angle between the stroke plane and longitudinal body axis, as it reveals the conditions under which the wing articulation works. In insects with low wingbeat frequencies this angle ranges from 70 to 100°, with an average of 90°. The ability to change this angle is restricted in insects with a low wingbeat frequencies, but increases in insects with higher wingbeat frequencies. Stellwaag (1916) was the first to notice that the honey bee varies the orientation of its stroke plane according to its direction of flight. In the same way, sawflies have a great deal of freedom of orientation of the stroke plane, while in Diptera any manoeuvre is accompanied by a change in the stroke plane angle relative to the longitudinal body axis. This angle is 'fixed' in cockroaches, mantids, scorpionflies, alderflies, and in all mayflies, and for the hindwings of stoneflies as well. The only way that mayflies and stoneflies can alter the stroke plane angle relative to the horizontal is to increase or decrease body angle. Neuroptera are exceptional. For example, in the green lacewing the change in stroke plane angle associated with the transition from forward to rising flight may be as much as 70° (Brodsky 1988).

In order to derive the vector of the resulting aerodynamic force and consequently the lift/thrust ratio, it is convenient to measure the stroke plane angle relative to flight direction, which coincides with the horizontal in the case of level forward flight (see Fig. 2.1). In insects with low wingbeat frequencies, the stroke plane angle relative to the horizontal is on average 60–80°. Insects with high wingbeat frequencies—fast fliers—have a more horizontal stroke plane. At low values of the stroke plane angle relative to the horizontal, an increased flight speed causes a decrease in the angle of attack of the wings to a lesser extent than at a steeper stroke plane angle, which can be illustrated by a simple diagram (Fig. 2.6). With a steep stroke plane angle an increase in flight speed and consequently of the horizontal component of the oncoming air current may considerably, if not completely, impede generation of aerodynamic forces because of the decrease in angle of attack of the wings. We might conclude from this that insects with a vertical stroke plane have to use a large angle of attack on the descending part of their wing trajectory. Furthermore, insects with a high wingbeat frequency and a sloping wing movement trajectory may obtain higher values of flight speed.

During manoeuvering, the wing on the inside of the turn side is held backward, and the one on the outside is held forwards. This only applies to the forewings in insects with two pairs of wings.

Body angle (pitch angle) is not directly linked to the work of the wings, but needs to be considered when discussing the nature of the stroke. During flight, body angle (Fig. 2.1) varies around an average value because of the oscillations in lift: at the end of the downstroke body angle increases, and at the end of the upstroke it decreases. During flight many insects hold

their bodies at a small average angle to the horizontal, which varies from 0 to 20° (most stoneflies, mantids, caddisflies, and many others). By contrast, mayfly subimagos, large stoneflies, green lacewings, scorpionflies, primitive caddisflies, aphids, small beetles, and some other insects are characterized by a steep (more than 50°) average body angle. Some of these insects (e.g. scorpionflies, cockroaches, and primitive caddisflies) possess limited capabilities for altering the flight speed, unlike mayflies, stoneflies, and other insects which fly with their bodies near horizontal. Very occasionally, body angle is negative (as in Corduliid dragonflies and some species of Rhopalocera). In general, insects with a high wingbeat frequency, apart from beetles, fly at a low body angle relative to the horizontal.

The *trajectories* described by the wing tip in flight are extraordinarily diverse (Fig. 2.7). Detailed analysis reveals well known features such as figures of eight or zero as well as combinations of these. The simple 0 shape is found, for example, in the alderfly (Fig. 2.7d), in the forewings of the brown lacewing (Fig. 2.7f), and in the sisyrid lacewing (Fig. 2.7g). The wing trajectories of other insects form a series of loops and are more complicated in shape—as in the hindwings of the sisyrid and brown lacewings, and in the scorpionfly (Fig. 2.7e)—those insects whose hindwing stroke angle is larger than that of the forewings.

The angle at which the airflow encounters the wing, the *aerodynamic angle of attack*, depends on the ratio between the flow velocity vector caused by the horizontal shift of the insect and the vector of relative wind caused by the flapping movement of the wing (see Fig. 2.6). To measure the aerodynamic angle of attack at a given point of the stroke, one needs to know how the angle of wing pitch alters relative to all points on the trajectory. The latter angle is often named the *geometrical* angle of attack. A number of investigators have analysed the wingstroke of insects by assuming a sequence of steady-state positions of the wing, and have tried to measure the geometric angle of attack at given points in the stroke. However, as the insect wing operates under unsteady flow conditions, the validity of making such precise calculations is questionable.

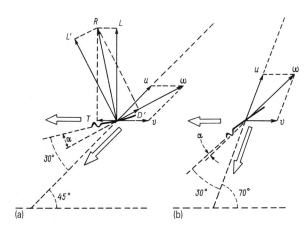

Fig. 2.6 Force and velocity systems of the wing during horizontal flapping flight at speed *v* with equal angles (30°) between the wing plane and the downstroke path when the stroke plane angle is small (a) and when it is large (b). Open arrows indicate the direction of the insect and its wing motion.

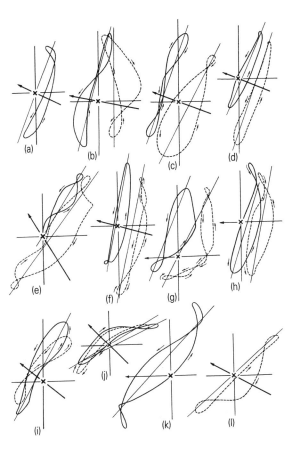

Fig. 2.7 The trajectories described by the wing tip during the forward flapping flight of various insects. The arrow indicates the direction of the insect head, the cross is centred at the base of the left forewing, and the path of the hindwing tip is shown by a broken line.

(a) Mayfly *Ephemera vulgata*; (b) stonefly *Isoptena serricornis*; (c) stonefly *Taeniopteryx nebulosa*; (d) alderfly *Sialis morio*; (e) scorpionfly *Panorpa communis*; (f) brown lacewing *Hemerobius simulans*; (g) sisyrid lacewing *Sisyra fuscata*; (h) caddisfly *Agrypnia obsoleta*; (i) planthopper *Cixius* sp.; (j) sawfly *Rhogogaster viridis*; (k) blow fly *Calliphora erythrocephala*; (l) beetle *Megopis* sp. (k) After Nachtigall (1966); (l) after Schneider (1980).

2.2.2 Characteristic wing movements

Recent interest in the uneven nature of insect flight has led to the identification of a great variety of wing movements which previously escaped notice. Some of these movements are linked to the alteration of the wing surface shape at particular moments in the stroke, and will be discussed in detail later on. This section focuses on movements which may be explained from a kinematic viewpoint, and which have been observed repeatedly in the flight of a number of species.

The *dorsal surface wing clap* occurs at the end of the upstroke—a movement to which attention was brought by Weis-Fogh (1973) when describing a new mechanism of lift generation in the tiny chalcid wasp *Encarsia formosa*. The same clap, in which the entire dorsal surfaces of the wings are clapped together, is observed in the Psocoptera, small Neuroptera, and Lepidoptera. However, in most cases only the apices of the wings are clapped together at the end of the upstroke. This is most typical in small insects such as leafhoppers, booklice, and relatively primitive ones (scorpionflies, alderflies, caddisflies, and lacewings). In large stoneflies, cockroaches, and mantids the clap only occurs in the hindwings (see Fig. 2.3, 1–3). The dorsal surface wing clap is not found in the Ephemeroptera or Diptera, and in the Hymenoptera it results from abrupt pronation of the wings, during which the posterior parts of the wings just touch (Fig. 2.8, 4).

The *ventral surface wing clap* which occurs at the end of the downstroke is obstructed by the thorax, which is why it is less common than the dorsal one. It has been observed in insects from a range of orders: the beetle *Donacia aquatica*, the stonefly *Chloroperla apicalis*, the caddisfly *Mystacides longicornis*, and the butterfly *Pieris napi*. Using an airflow visualization technique, my colleagues and I have managed to follow the path of visualizing particles during the clap. In the caddisfly, the clap at the end of the downstroke, like that at the end of the upstroke, is accompanied by the same phenomena: when the wing tips are flung open, air rushes into the vacuum created by the separating wings and causes a vortex whose direction of rotation

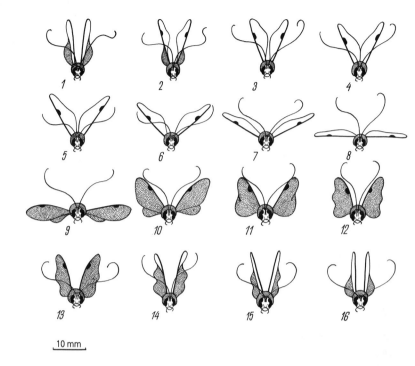

Fig. 2.8 Flight of the ichneumon *Ophion* sp. (front view). Consecutive film tracings (1–16) from a single wingbeat. Stroke parameters (see list of abbreviations, p.xi): $n = 90$ Hz, $A = 90°$, $b = 45°$, $B = 0°$. The undersurface of the wings is shaded.

|___10 mm___|

coincides with that of the starting one (Fig. 2.9a). In the butterfly *Pieris napi*, wing contact at the end of the downstroke is more complete than in the caddisfly, and is accompanied by a powerful flow which is projected upward over the head and then backward (Fig. 2.9b). At that moment an abrupt increase in lift occurs which is evident from the elevated thorax flight posture of this butterfly.

Pronation (=fling) is an important element of the wing motion of any insect. Depression of the wings at the end of upstroke during which the wings rotate about their longitudinal axes with the costal margins facing down (see Figs 2.8, 1–5 and 2.10, 1–3) begins with pronation. At the end of pronation the wings stop for a moment, supinate slightly (that is, rotate with their anterior margins upward (Figs 2.8, 7 and 2.10, 4 and 5)), and only then start to move down.

The term *'quivering'* epitomizes frequent small waves of movement which run over the wing membrane. It has been observed at the end of the downstroke in the crane fly *Tipula paludosa*

(Brodsky 1985), and also during the first half of the downstroke in the flies *Phormia* and *Calliphora* (Nachtigall 1980).

Flutter describes low-amplitude, high-frequency oscillations of the wing tip. These oscillations have been observed during the last third of the upstroke in the caddisfly *Limnephilus stigma* and the antlion *Myrmeleon formicarius* (Brodsky 1985), and presumably result from stalling of the leading edge. Flutter is occasionally observed in flight experiments with tethered insects in still air.

Flick is a vigorous supinational movement which occurs at the start of the upstroke, accompanied by rapid bending and then the immediate straightening of the wing tip (Fig. 2.9c). It is typical of the forewings of stoneflies, but is rare in other insects in which the wing is twisted less.

A wave running along the posterior margin of the wing is observed during the flight of many insects. In the moth *Stathmopoda pedella*, an undulating disturbance of increasing amplitude

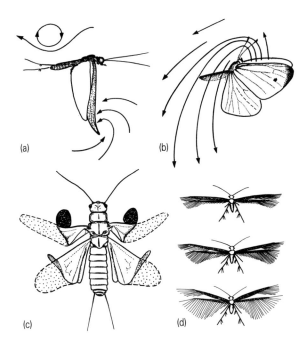

Fig. 2.9 Some characteristic wing movements. (a) Clap with the ventral sides of the wings at the end of the downstroke in the caddisfly *Mystacides longicornis*. The arrows indicate the direction of airflow in the region near the clap. (b) Clap with the ventral sides of the wings in the pierid butterfly *Pieris napi*. (c) Flick of the forewing tips in the stonefly *Isogenus nubecula*. (d) Wave passing along the rear margin of the wing in the moth *Stathmopoda pedella*.

passes from the wing tip to its base along the long hairs of the posterior margin during supination (Fig. 2.9d).

2.2.3 A three-dimensional assessment of the wing stroke

A great deal of data has been obtained on wing motion during the stroke. When investigating the wing motion of insects, one is compelled (because of the limitations of one's instruments) to consider one of three reciprocally-perpendicular planes: sagittal, frontal, and transverse. In each of these planes the wing motion

looks like flapping—coming to a stop, and then accelerating. If, for instance, wing motion is being analysed in the sagittal plane, the investigator observes that the wing stops at the extremes of the up- and downstrokes. These points separate the descending and ascending paths of the wing motion. The fact that the wing does not stop at the ends of up- and downstrokes but continues its motion, albeit in a different plane (rotating on its longitudinal axis), is not observed. Clearly, accumulating data on wing motion in each of these separate planes can hardly lead us to an accurate assessment of the motion of the wings in space. One has to go beyond the limits of a traditional one-dimensional analysis of motion, and try to make a three-dimensional assessment.

Up to now there has only been one attempt to analyse the wing motion of insects in terms of a rigid aerofoil with one fixed point, by Zarnack (1972), who studied the motion of the forewing of the migratory locust in conditions close to free flight. Zarnack used high-speed film to investigate the time (t) dependence of the Eulerian angles $\psi(t)$, $\varphi(t)$, and $\theta(t)$ which determine the position of the wing coordinate system in relation to the stationary coordinate system. Fig. 2.11 shows the curves obtained from values of Eulerian angles calculated by Zarnack at discrete moments in the stroke. ψ, φ, and θ are rather characteristic functions of time: ψ and φ are close to linear, while θ is close to constant. It is worth considering what form the wing motion would take if φ and ψ were linear ($\varphi = \varphi_0 + \omega t$, $\psi = \psi_0 - \omega t$) and $\theta = \theta_0$ were constant. Such a case of movement of a solid body with one fixed point is well known in theoretical mechanics as retrograde regular precession—the angular velocities of precession and of proper rotation are constant and bear opposite signs. In fact, this is a particular case of retrograde precession in which the periods of precessional and proper rotations are equal. An aerofoil which moves like this has a wingbeat frequency of $n = \omega/2\pi$, and the same phases of movement as the wing of a real insect. Pronation and supination in this case appear to be intrinsic parts of the motion, and the wing tip describes a figure-of-eight in space, with its shape varying both with the rotational phase shift and with small nutational oscillations (deviations of θ from constant during the stroke).

The motion discussed above can be expressed in relatively simple mathematical terms and can be defined by a small number of parameters (θ_0; $\varepsilon_0 = |\psi_0| - |\varphi_0|$, the phase shift of proper and precessional rotations; z_0, the unit vector of the precessional axis; ω, the angular velocity of precession which is equal to the angular velocity of proper rotation and wingbeat frequency). These parameters relate to those used in the traditional description of the wing motion of insects as follows: the angle θ_0 determines half of the stroke angle; the inclination of the precessional axis z_0 defines various angles of tilt of the stroke plane (both relative to the horizontal and to the vertical); and the parameter ε_0 allows the geometrical angles of attack of the wing to be altered with almost constant stroke angle and stroke plane angle.

Let us define the main kinematic features of the motion we are considering. Wing motion during the stroke appears to be a combination of two rotational movements: proper and precessional. The axis of the moment of rotation gradually changes its position in space, while the value of the momentary angular velocity remains

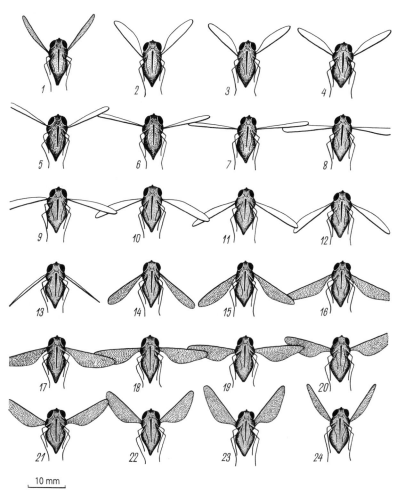

10 mm

Fig. 2.10 Flight of the nemestrinid *Nemestrinus capito* (front view); consecutive film tracings (1–24) from a single wingbeat. Stroke parameters (see list of abbreviations, p.xi): $n = 143$ Hz, $A = 90°$, $b = 30°$, $B = 60°$. The undersurface of the wings is shaded.

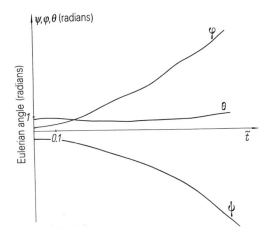

Fig. 2.11 Variation in the Eulerian angles ψ (*t*), φ (*t*), and θ (*t*) over time in the locust forewing averaged over the stroke period. The wing movement begins with the downstroke and continues a little further than one stroke cycle.

constant. Its projections on the movable axes p, q, and r are equal to:

$$p = -\omega\sin\theta_0\sin\varphi,$$
$$q = -\omega\sin\theta_0\cos\varphi,$$
$$r = \omega(1 - \cos\theta_0).$$

We know that twice the value of kinetic energy is equal to $Ap^2 + Bq^2 + Cr^2$, where A, B, and C are the moments of inertia which correspond with each of the major axes. If we equalize the moments A and B relative to the major axes of inertia in the wing plane, the motion may proceed even when the entire kinetic energy is stored. It is clear that in this case even the term 'flapping flight' is inaccurate.

The movement of the wing we are discussing includes flapping (up-and-down strokes). The position of the wing in space during the stroke is one-sided relative to the sagittal plane, which passes through the longitudinal axis of the insect's body. This follows from the Eulerian formulae: at $\varepsilon_0 < \pi/4$, the values of the y coordinate of the wing tip have the same sign. One of the wing edges (the leading edge) is always upstream of the other (the trailing edge).

However, the wing motion can hardly be defined as 'flapping' in the generally accepted sense of coming to a stop and then accelerating at the ends of the up-and downstrokes, since the momentary angular velocity remains constant, the position of the momentary rotational axis changes steadily, and kinetic energy may be stored in the case of corresponding distribution of masses along the wing. When one considers both the features described above and the mechanism by which the strokes are produced, which differs in principle from those used by insects, it is convenient to designate such wing motion as 'quasi-flapping'.

The quasi-flapping wing motion can be simulated easily with a device made up of two drives (Fig. 2.12). By changing the characteristics of the proper and precessional rotations, we can restore

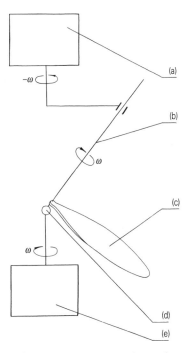

Fig. 2.12 Schematic representation of a device providing a quasi-flapping wing motion. (a) Drive providing the precessional rotation of the wing with angular velocity − ω; (b) axis of wing rotation; (c) wing; (d) hinge; (e) drive providing the rotation of wing axis with angular velocity ω.

the spatial picture of wing motion, as characterized by certain stroke parameters: stroke angle, wingbeat frequency, stroke plane angle relative to the horizontal and vertical, and the angle of attack of the wing during the up- and downstrokes. The proposed model saves

students from the time-consuming process of reconstructing the spatial picture of wing movement during strokes (Zarnack has only achieved this once, for the migratory locust) and allows them to make spatial aerodynamic calculations for insects with different stroke parameters.

2.3 Wing deformation during flight

Insect wings undergo strong deformation during flight, the character of the deformation varying between species. Depending on how all or part of the wing plane moves, typical wing deformation may occur in members of an order, family, or group of families.

The character of the wing deformation depends on wing structure—on the elements which provide support and permit deformation. In addition, individual sections of the wing can change position during flight without affecting the attitude of the wing plane as a whole; these individual sections break down the airflow into separate currents and are responsible for their direction of flow down the wing. As a consequence of their roles in flight, the various elements have differentiated into three functional groups: supporting, aerodynamic, and deformable structures.

2.3.1 Supporting structures

The supporting elements or structures which provide the rigidity of an insect wing comprise thickened veins; corrugation of the wing plane serves the same purpose. As well as along the main mechanical axis, other structures in the wings give support to the posterior area of the wing plane. This function is usually performed by the postcubital vein on the clavus. A further adaptation worth mentioning is the development of rigid fields in the wing delimited by supporting elements, as in the base of the clavus of stoneflies or between the radius and the radial sector in alderflies (Fig. 2.13a). Such supporting fields function as single strong 'veins'.

In many cases supporting structures control deformation of the wing, limiting its extent or the location of folds to certain areas of the

wing. Such is the role of an 'x' shape at the wing tip in stoneflies, and of the veins which determine the position of the creases in caddisfly and alderfly wings. It should be noted that in most cases the supporting veins differ very little from other veins in network of veins, and their roles can be most easily identified by their weak deformability throughout a stroke cycle.

2.3.2 Patterns of wing deformation

The deformable elements are those structures that can be conveniently grouped on the basis of patterns of wing membrane deformation.

Longitudinal bending of the wing plane during the stroke occurs along particular lines (Fig. 2.13a). The lines in which supination is concentrated are usually fixed, and correspond with some of the concave folds. These folds bound posteriorly the area of the wing which rotates about the wing's longitudinal axis such that this area's anterior margin elevates while the posterior one depresses; the fold itself points downward. The fold at which the maximum deformation occurs can be further designated the main supination line; supplementary folds, which deform less, constitute accessory supination lines.

In the forewings, the main supination line is the cubital fold, and the accessory supination line is the radial sector fold. In the hindwings of stoneflies and some caddisflies, the fold of the radial sector is the main supination line.

Lines of pronation are frequently less distinct than those of supination in insect wings. This is because pronation often consists of straightening the wing along the lines of supination. Sometimes the forewings bend along the jugal or another fold. In cases where the

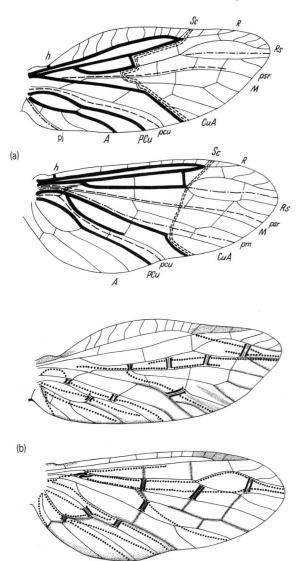

(a)

(b)

Fig. 2.13 Distribution of the supporting, aerodynamic, and deformable elements of the fore- and hindwings of the alderfly *Sialis morio*. (a) Supporting elements are shown with a solid line, lines of pronation are shown with a dot-dash line, lines of supination with a dashed line, and a line of tip deflection (transverse fold) with a double broken line. (b) Other lines of wing deformation are shown with a dotted line, short double solid lines indicate the positions of the main wing hinges, and aerodynamic elements are shaded.

Fig. 2.14 The owlet moth *Phlogophora meticulosa* in free flight, manoeuvering among plants. The wing tips bend downwards during the upstroke, forming an angularly configured sharp flexion line (after Dalton 1975).

wing becomes convex during the downstroke, either bending does not occur along any particular line, or it occurs along the fold in the medial field (broad-winged Lepidoptera and some Diptera (Orthorrhapha and Syrphidae)). In some caddisflies and alderflies, the hindwings pronate along the medial fold (Fig. 2.13a). The development of the 'Z-profile' described by Jensen (1956) in the desert locust is associated with raising of the radius, giving rise to apparent bending along the convex fold. The radial sector fold, which presumably forms part of the Z-profile, does not contribute to pronation in any species studied, and is always concave.

Sometimes insect wings show a particular type of pronation such as the fling. In this case the wings separate while bending along a line which moves backwards with time. This mechanism occurs more frequently during pronation of the hindwings; in this case, pronation usually starts with bending along the postcubital (caddisflies) or the anal (stoneflies) fold.

Deflection of the wing tip occurs in many insects. Sometimes ventral bending may take place over the entire distal part of the span by simple elastic deflection (see Fig. 2.3, 10–12), but it is frequently localized along a particular transverse line curved in a semi-circle or angled diagonally (Fig. 2.14). The transverse fold which allows this deformation nearly always starts from the apex of *Sc*, and may cross, and hence interrupt, the supporting veins. Membranous fields lie along both sides of the radial vein, and the radial vein itself is presumably constructed so as to allow repeated bending without losing stiffness. The transverse fold is often in effect a boundary between the basal supporting area and a distal deformable area (Fig. 2.13a), allowing the wing to bend ventrally along a sharp diagonal flexion line.

As has been observed by high-speed filming, deflection of the tip develops at the beginning of wing deceleration, in the lower part of the wing path, and ends at the beginning of upstroke. This deflection of the tip obviously results from the action of inertia forces during deceleration of the wing. Long wings undergo stronger deflection of the tip because the forces of inertia are more substantial. The pterostigma, which is distal from the transverse fold, regulates the extent of deflection. As an inertial regulator, it is located such that it provides effective deflection of the wing tip without impeding maintenance of the primary position by the tip.

The absence of tip deflection at the end of the upstroke may be explained by the fact that wing deceleration and acceleration are provided by the elastic properties of the thorax at this stage; besides, the wings stop by clapping against each other.

Torsion is observed in many relatively primitive insects (cockroaches, mantids, stoneflies, alderflies, and others); it develops at the end of the downstroke and persists throughout the upward movement of the wing. Torsion is expressed to a greater extent in wings which lack an anal fan and supinate along the cubital fold. The role of the clavus in the development of torsional deformation during the wing upstroke is of especial interest. Rapid twisting occurs at the wing base, including the remigium and the clavus. Then, at the beginning of supination, a stiffening rib appears along the cubital fold, as a result of which the twisting area shifts to the wing tip, and the curvature of the clavus alters. During the upstroke the clavus regulates

the degree of wing twisting, damping its propagation through the wing at the basal areas of the wing plane. The degree of twisting depends on the shape of the clavus and the length of the cubital fold. Thus in cockroaches with a short, curved cubital fold and rounded clavus, twisting during the upstroke is considerable. In stoneflies with a 'classic' wedge-shaped clavus, the deformation is less. Weakest torsion is observed in caddisflies which possess a very long cubital fold and an oblong clavus.

Wing torsion may occur other than at the end of the downstroke. In many insects the wing undergoes pronational twisting as it moves downward (Fig. 2.15) and a consequent diminution in angle of attack from base to tip. Wootton (1981) attributed this deformation to backward displacement of the centre of aerodynamic pressure relative to the rotational axis of the wing (Fig. 2.16).

Camber (a convex profile) in insect wings is observed throughout the downstroke, whereas during the upstroke the profile is concave. This occurs in some insect wings—Orthoptera, Lepidoptera, and others—whereas in Neuroptera, Mecoptera, and Odonata the wing profile remains almost flat throughout the up- and downstrokes. The concave profile of the wing during the upstroke is maintained by strong wing supination (Fig. 2.17), and serves to reduce negative lift at this part of the stroke. The role of the convex profile during the downstroke is less clear, but from classical aerodynamics it is well known that a cambered wing produces greater lift than a flat one.

The wave passing along the posterior margin of a wing has already been mentioned in this chapter. In many insects, at different moments of the stroke, an undulation passes along the flexible posterior margin. During supination and pronation an angular disturbance occurs near the wing tip, and then, like a running wave, passes to the body (Fig. 2.18, 1–12); from there, because of the elasticity of the wing structures, it may move apically (Fig. 2.17, 13–24). Frequent undulating disturbances and their rapid passage through the wing cause additional high-frequency oscillations which influence the pitch of the sound produced by a flying insect.

2.3.3 Aerodynamic wing structures

The aerodynamic elements, unlike all other structures, are only concerned with the distribution of air currents over the wing surface. During flight, separate parts of the wing margin deflect, and various deformations of the wing plan occur which affect the wing-airflow interaction. The aerodynamic structures alter the local angle of attack of the wing, break up the airflow close to the surface into currents, and regulate the airflow off the wing.

The grooves and ridges which exist on the wing or which appear at some stage of the stroke cycle can be classified into four groups. Some of these grooves and ridges are located

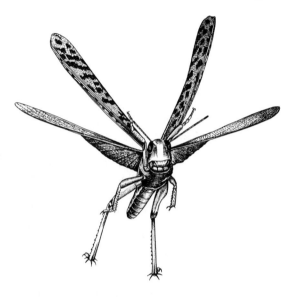

Fig. 2.15 At take-off the desert locust, *Schistocerca gregaria* jumps and then flaps its wings downward. During the downstroke both fore- and hindwings undergo strong pronational twisting (after Dalton 1975).

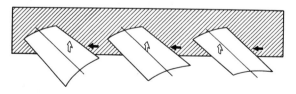

Fig. 2.16 Effect of the relative positions of the centre of aerodynamic pressure and the torsional axis (the line on the wing) on the aerodynamic torsion of a fixed wing. Solid arrows indicate the direction of the relative wind and open arrows the lift at the centre of pressure (after Wootton 1981).

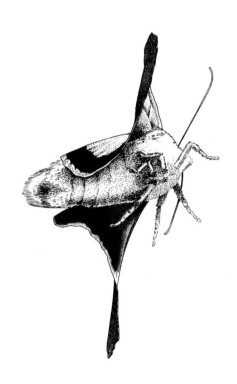

Fig. 2.17 When disturbed, the underwing *Lampra fimbriata* quickly flies off in a erratic manner. During the upstroke, the concave profile of the wing is maintained by strong supination. The concave profile is superposed on the screw-like torsion of the wing which is beautifully evident as the underwing changes course (after Dalton 1975).

in the marginal area, perpendicular to the wing margin, and broaden gradually up to the edge (see Fig. 2.13b). These grooves are formed during the fastest movements; either in the middle of a stroke or on a bent part of the wing when it is rapidly unbending, and are a natural occurrence for free posterior and terminal edges. During the downstroke concave grooves occur on the wing, not convex ones as would be expected. These are presumably formed actively by the insect.

The grooves and ridges which are grouped in the middle of the wing are oriented longitudinally. During the downstroke, flexures along the radial fold and the fold of the radial sector become clearly distinct (see Fig. 2.13b). Along these temporary grooves flow base-to-tip currents as a result of centrifugal forces.

During the upstroke grooves confluent with the cubital fold are formed. These may occur on the clavus (in caddisflies) and also in the posterior part of the remigium (in stoneflies). The formation of the latter is actively assisted by the transverse veins between M, CuA, and CuP—the 'combs'. These structures are most distinct during the upstroke when the cubital fold constitutes a powerful collector for the air currents which stream down from the supinated wing.

The fourth group of aerodynamic structures is concerned with the wing tip. In caddisfly wings, strongly curved ridges form near the tip of the discoidal cell at the end of the upstroke. These presumably serve in the formation of a vortex in the apical part of wing at the end of the upstroke and at the start of downstroke. The upward bending of the wing tips at the end of upstroke in stoneflies, and the ridges produced by the convex transverse veins in alderflies and scorpionflies probably fulfil the same function.

Fig. 2.18 Consecutive film tracings (1–24) from the upstroke of the left wing of the crane fly *Cylindrotoma distinctissima*, as seen from above. Propagation of a torsion wave from tip to base (1–12) brings the wing into a supinated state. Propagation of torsion wave in the reverse direction, i.e. from base to tip (13–24), brings the wing into a slightly pronated state. Note that the anal lobe is the last part to bend (8, 9) beneath the wing plane during supination, and the first to straighten (15–17) during pronation, thus storing and releasing energy for wave propagation.

3

The aerodynamics of insect flight

The question of how aerodynamic forces are generated has been and continues to be the most basic question in the study of insect flight. Although it was addressed by the very first investigators of insect flight, Demoll (1918) and Magnan (1934), Holst (1943) was the first to prove experimentally that the previous concept of insect flight as 'rowing' does not correspond to how the flapping wing actually works. In contrast to previous over-simplified and imprecise notions about air being 'thrown off' by the wing, the quasi-steady approach enabled aerodynamic processes to be considered as a series of stationary conditions which could be quantified using classical aerodynamic methods. An investigation of the flight of the desert locust by Weis-Fogh and Jensen (1956) obtained a good match between theoretical data and measured values of aerodynamics forces. In spite of general acceptance of the quasi-steady approach, nobody else has managed to obtain agreement between theoretical and real forces, which implies the existence of additional aerodynamic forces. In addition, many authors have emphasized the non-stationary nature of the forces created during insect flight, and the unusual behaviour of the airflow in the insect's wake.

3.1 The physical basis of flight

Locomotion in continuous media like water or air differs so much from motion at the boundary between two media that the nature of the forces which make it possible deserves special attention. A continuous medium is one in which there is nothing to 'lean on' or 'push off'. There are two principal forms of flow in a viscous fluid such as water or air: laminar or foliated flow, in which the streamlines are parallel to each other; and turbulent flow, in which the fluid is mixed. Neither velocity nor pressure in turbulent flows are constant at any given point and they occur in irregular, high-frequency pulses.

The surface of a body moving in a continuous medium experiences both normal and tangential forces. Their resultant is the total mean drag force on the moving body, though its vector does not always coincide with the direction of motion. The projection of the total body drag on the x-axis in a velocity-oriented coordinate system (in which the x-axis lies along the velocity vector, and the y- and z-axes are perpendicular to it in horizontal and vertical planes respectively) is the frontal drag, and its projection on the z-axis is the lifting force when z is positive, and either a sinking force (in water) or a negative lifting force (in air) when z is negative. The frontal drag comprises friction drag and pressure (or form) drag (the pressure difference between the front and rear of the body). The values of the frontal drag (D) and the lifting force (L) are given by the formulae:

$$D = C_D \frac{\rho V^2}{2} S \text{ and}$$

$$L = C_L \frac{\rho V^2}{2} S,$$

where S is a characteristic area of the body, ρ is the density of the medium, V is the flight

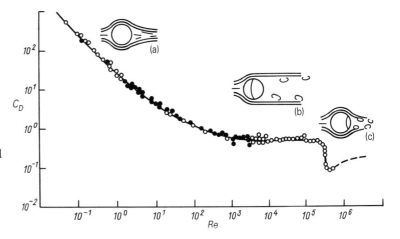

Fig. 3.1 Dependence of the drag coefficient of a sphere on Reynolds number. Schematized diagrams over the curve show the flow around the sphere in three different Reynolds numbers' regions: (a) for crawling movements, and at (b) subcritical and (c) critical Reynolds numbers.

speed, and C_D and C_L are non-dimensional coefficients of frontal drag and lift respectively. C_D and C_L depend first and foremost upon the inter-correlation between the inertial forces (manifested as pressure on the body surface) and the frictional forces, which are determined by the viscosity of the fluid, and upon the shape of the moving body, and its orientation. These inter-correlated forces form a dimensionless number, whose value is equal for equal flows. In other words, we are considering the scale criterion that allows us to compare the motion of different objects with a variety of characteristics. This number is called the Reynolds number (Re):

$$Re = \frac{\text{Inertial force}}{\text{Frictional force}} = \frac{\rho V^2 / l}{\mu V / l^2} = \frac{\rho V l}{\mu} = \frac{V l}{\nu},$$

where μ is the viscosity of the medium, ν is the kinematic viscosity of the medium, and l is a characteristic linear dimension of the body.

Figure 3.1 illustrates the effect of the Reynolds number on the frontal drag coefficient of a sphere. When the Reynolds number is small (up to 1) the body drag depends only on frictional forces, and the orientation of the body in space is unimportant (Fig. 3.1a). In the next range of Reynolds number values, the laminar boundary layer[1] develops gradually along the contours of

the sphere and then breaks away from them. The broken-off boundary layer curls round to form two vortices which remain symmetrically behind the body for some time (Fig. 3.1b). A further increase in Reynolds number leads to the transition of the boundary layer into a turbulent state; shedding happens downstream along the contour, as a result of which the perturbed region behind the body narrows (Fig. 3.1c). This is to do with the decrease in pressure drag which, even with increased friction drag (due to transition of the boundary layer into a turbulent state), results in a reduction in total frontal drag. Reynolds numbers in this range are called 'critical' values. At supercritical Re values, the flow in the boundary layer is partially turbulent and vortices form in two rows behind the body, in a chequer-board pattern. With further increases in Re a strongly pronounced turbulent wake becomes apparent behind the body.

Fig 3.2 shows the average velocity of animals in a continuous medium as a function of Reynolds number. The range of Re for biological objects is considerable: from 10^{-6} (for

---

[1] The boundary layer is the region close to the boundary between the fluid medium and a solid body, where fluid is slowed by the solid body due to friction.

bacteria) up to 10^7 (for a large cetacean). This continuum begins with small organisms whose locomotion is wholly dependent on viscosity and ends with large swimming animals, whose loco-motion is mainly influenced by inertial forces. As soon as an infusorian stops moving its cilia, it immediately comes to a halt. But a fish only needs to beat its tail once to glide forward for a comparatively long time. When a thrips stops moving its wings, it comes to a halt, while a bird continues to move forward through inertia.

As the significance of inertial forces increases with increases in Re, the lifting force becomes more important in animal locomotion. At $Re \simeq 10^3$, the lifting force is three or more times the frontal drag. Thus the ratio of lift to frontal drag for the wing of a blow fly *Calliphora* is 3:1, while for larger insects it is even greater. In animals in which inertial forces are greater than frictional ones, locomotion is based mainly on lifting forces.

Inertial forces give way to frictional forces when Reynolds numbers are small. In this situation, use of lifting forces is impossible. What principle animals use for locomotion in this zone is not yet quite clear, as the experimental study of moving animals under these conditions encounters severe practical difficulties. This is why the famous investigator of insect flight Weis-Fogh called this range of Reynolds numbers (10^0–10^2) the 'twilight zone'. Frictional forces take over from inertial ones, but neither their extent nor how they create a propulsive force is known.

When inertial forces prevail during locomo-tion, Newton's third law, which states that for every action there is an equal and opposite reaction, is obeyed. This means that if an animal propels itself through a fluid, be it water or air, it requires momentum as thrust to over-come frontal drag and lift to overcome body weight. Therefore, equal and opposite momen-tum must be generated in the fluid flow form-ing the wake of the animal. This wake (a so-called 'hydrodynamic trace', or 'wake zone')

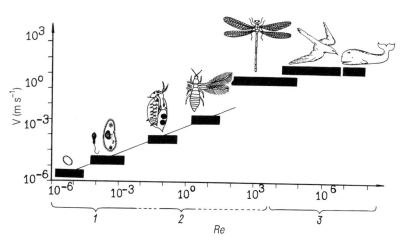

Fig. 3.2 Speed of travel of biological objects as a function of Reynolds number. Figures below the Reynolds number axis show regions of different viscosity-inertial effects: 1, pure viscosity effects; 2, both effects; 3, pure inertial effects (after Nachtigall 1981).

consists of separated flow particles rolled up into vortices. Various turbulent perturbations from the oncoming airflow also pass into the wake. Non-steady propulsive agents (such as fins or wings) cause the wake zone to be filled with intensive vortices, whose disposition is strictly regular in the absence of turbulence. If propulsive forces are created, the vortices arrange themselves behind the oscillating body in two rows, rotating alternately in a chequer-board pattern. This sort of wake is reminiscent of the so-called 'Kármán's vortex street'

(Kármán and Burgers 1935) which forms behind a badly streamlined body (Fig 3.1b) except that in this case the vortices rotate in the opposite direction.

In a free fluid medium—that is, at a distance from a solid surface—vortex rings are the only structures capable of retaining the momentum passed to the medium by the organs acting upon it. Consequently, the hydroaerodynamic wake of any animal, be it a swimming fish, a flying insect, or a jetting cephalopod mollusc, must consist of vortex rings.

3.2 The range of Reynolds numbers of flying insects

The Reynolds number is the best indicator of the type of force generation that is used in the flight of particular groups of animals. Defining the range of Reynolds numbers over which insect flight occurs is not easy, as the conditions of airflow around the wings change throughout the whole wingbeat cycle. The flapping movements of the wing are based on a combination of progressive and rotational (angular) oscillations. The phase offset (phase angle) between these oscillations may vary. The wing rotates around its longitudinal axis (its angular oscillation) at the extremes of the flap, when the wing swings round to continue the motion with a new angle of attack. The rotational and the progressive oscillations of a wing both occur over a range of velocities. Consequently, the Reynolds number which characterizes a given wing varies with time and also varies topographically. In defining Re for flapping wings, the maximum (or sometimes the mean) length of the wing chord is usually used as the characteristic dimension, and its motion velocity at the middle of a downstroke (when the wing moves uniformly) is defined as the vector sum of the flapping and flight velocities. Using this simplified method of Re definition, we obtain the following Re values for two insect species of approximately equal size but different kinematic parameters and wing form: the mayfly *Ephemera vulgata*, 1073; the crane fly *Tipula paludosa*, 1130. It can be seen that the values of Re for both species are similar, even though the crane fly's wingbeat

frequency (58) is twice that of the mayfly (25) so that one would expect that their aerodynamic characteristics would differ.

In view of the difficulties described above, it is more convenient to evaluate the Reynolds numbers of flying insects on the basis of the airflow conditions around the body. Using the vertical axis of the frontal body projection as the characteristic linear dimension, and the typical flight speeds of several species, Brodsky and Ivanov (1973) showed that the flight of comparatively large and fast-flying insects may occur at critical Reynolds numbers (Fig. 3.1c). A similar result was obtained by Newman *et al.* (1977) for a dragonfly wing ($Re \simeq 10^4$) fixed in a gliding position and exposed to a flow with a low degree of turbulence. The fact that insects can fly in the critical Reynolds number range is of great interest, for it implies that in this region insects have an effective mechanism for creating turbulence in the boundary layer, thus decreasing the pressure drag.

This sort of information enables us to obtain some idea of the range of Reynolds numbers in which insect flight does take place. Because of the measurement problems described above, differences in rates of less than one order of magnitude cannot be relied upon. In general, Reynolds numbers range from 10^0 (for the smallest insects, with fringed wings) up to 10^4 (for large beetles, moths and butterflies, and anisopterous Odonata). The range is clearly wide enough for us to consider that the variation in frictional and inertial forces is real, i.e. that these insects have different aerodynamic flight characteristics.

The relationship between the ratio C_L/C_D and the angle of attack (an aerodynamic quality of wing) gives some indication of the aerodynamic characteristics of flight. This relationship depends not only on the profile and plan of the wing, but also on the Reynolds number. The polar diagrams of various insect wings shown in Fig. 3.3 fall into three categories. At high Reynolds numbers, the highest value of C_L/C_D occurs at low values of angle of attack (Fig. 3.3a). Thus, the C_L/C_D ratio for the hindwing (Fig. 3.3b) of a desert locust ($Re = 2000$) is 8:1 at $\alpha = 10°$, but at $\alpha = 20°$ it drops to 5:1, and at the critical value ($\alpha = 25°$), the airflow stalls. Hence, the most economical and effective work regime for such a wing is for it to move with a small angle of attack. The second category of polars

(Fig. 3.3c–e) shows the absence of a sharp stall. At $Re \simeq 1000$, the C_L/C_D ratio is considerably smaller than at $Re = 2000$, and it reaches its highest value (3:1) at $\alpha = 10°$. With an increased angle of attack this ratio is only 1.5:1, and does not change a great deal even at very high values of angle of attack. The polar diagram (Fig. 3.3f) of the wing of a fruit fly *Drosophila virilis* (in which $Re = 200$) falls into the third category. In this case the efficiency of the work of the wing is practically independent of the angle of attack, and C_L/C_D is equal to 1:1. In other words, in *Drosophila* flight frictional forces are as important as are inertial ones. Thus all winged insects can be divided into three groups depending on the relative contributions of inertial and frictional forces in the creation of propulsive force. The larger insects, such as desert locust, anisopterous Odonata, sphinx moths, etc. belong to the first group. Here inertial forces obviously prevail over frictional ones. Middle-sized insects belong to the second group (mayflies, many flies, zygopterous Odonata, etc.). During these insects' flight inertial forces take over from frictional forces, but not significantly; the main differences in airflow around the wing are probably caused by differences in wingbeat frequency.

The Strouhal number (Sr) is a dimensionless number relating main flow velocity and the velocity of periodical (oscillating) motion:

$$Sr = nl/V,$$

where n is wingbeat frequency, l is wing length, and V is flight speed. However, this number in its classical form is unsuitable for describing insect flight since it does not take into account the amplitude of the wingbeat. It is preferable to use another dimensionless number named the 'reduced', or non-dimensional frequency $k = 2nAr/V$, where A is the wingstroke amplitude, and r is the distance from the wing base to the chord in question. It is easy to see that at $V = 0$ (hovering flight) k tends to infinity, and that at $n = 0$ k is equal to zero, i.e. the insect glides without flapping. As the main role in vortex formation is played by the wing tip, it is convenient to modify the reduced frequency to $k = nS/V$, where S is the length of the path taken by the wing tip during a full flap.

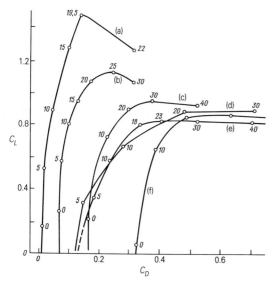

Fig. 3.3 Aerodynamic polars of an aeroplane and insect wings. (a) Aeroplane profile NACA 2409. (b) Locust hindwing (after Jensen 1956). (c) Forewing of the mayfly *Ephemera vulgata* (after Brodsky 1970*a*). (d) Wing of the dragonfly *Calopteryx splendens* (after Rudolph 1976). (e) Wing of the crane fly *Tipula* sp. (after Nachtigall 1977). (f) Wing of the fruit fly *Drosophila virilis* (after Vogel 1967*b*).

Comparing the mayfly *Ephemera vulgata* and the crane fly *Tipula paludosa* which have nearly equal Reynolds numbers, we obtain *k* values of 1 and 3 respectively. This implies that the periodical oscillations in airflow around the wing are three times more important for crane flies, which must have a significant effect on the aerodynamic characteristics of their flight.

3.3 Non-steady aerodynamic approaches

As the insect wing oscillates, it may accelerate or brake; and at the extreme points of the stroke, it rotates around its longitudinal axis, with a rotation velocity which is similarly irregular and which differs from that of the up-and-down movements. All this make it quite difficult, indeed almost impossible, to use principles of steady-state aerodynamics to explain the mechanism by which forces are created in flapping flight. Nevertheless, many people have clung to the traditional 'wing theory' of flapping flight, whose central concepts have been investigated theoretically and consolidated by numerous experimental data. In the description of flapping flight on the basis of a quasi-steady approach it is assumed that an insect wing is a thin plate round which the air flows at a constant velocity, so that force coefficients are constant for a given wingspan and over time, and that no aerodynamic interaction occurs between right and left wings.

At the centre of modern wing theory lies Chaplygin and Joukowsky's postulate[1], which states that the airflow of the upper and lower wing surfaces flow off the trailing edge of the wing. As soon as the wing starts to move, a vortex is formed on its trailing edge. This vortex grows rapidly until the motion of the fluid around the trailing edge ceases, i.e. until the air flows smoothly off both the upper and lower wing surfaces at the trailing edge (Fig. 3.4a (iii)). As soon as this happens, the vortex is shed and is carried away by the airflow. According to Thompson's theorem of circulation constancy, this shedding of the vortex induces a circulation around the wing, which might be considered a 'bound circulation' (Fig. 3.4a (ii)) whose direction of rotation is opposite to that of the starting vortex. By adding a circulating flow to that which would exist if no lift were produced (Fig. 3.4a (i)), the average air velocity increases over the upper surface and decreases over the lower one. This difference in velocities determines the distribution of pressure on the aerofoil (Fig. 3.4b), as is well known from classical aerodynamics, in which the value of lift per unit wingspan is defined by Joukowsky's theorem which states that $L = \rho V \Gamma$, where Γ is the circulation flow around the wing. The dependence of the lift and drag coefficients upon the angle of attack can be expressed by Lilienthal's polar, shown as a curve representing the vector of the full aerodynamic force as a function of changing angles of attack (Fig. 3.4c).

In the above approach, the airflow round a flapping wing is considered as a sequence of steady states; for each of such states changes in angle of attack and in oncoming flow velocity may be neglected. There are two moments in the flapping cycle that seem to satisfy conditions of a quasi-steady approach: the first corresponds to the main part of the descending branch of the trajectory, and the second corresponds to the lower third of the ascending branch (Fig. 3.4d). During the downstroke the wing generates lift and thrust, while in the upstroke, it generates thrust and negative lift. It is not known whether the wing creates any forces during other phases of the flapping cycle, because the processes going on at the upper and lower extremes of its trajectory can not be described using a quasi-steady state approach.

The problem of creating of aerodynamic forces by means of a flapping wing can be observed from another viewpoint. As a result of the

[1] Basic aerodynamic concepts discussed in this section may be found in texts such as Prandtl and Tietjens (1957), Mises (1959), Duncan *et al.* (1970), Fung (1969), and Bramwell (1976).

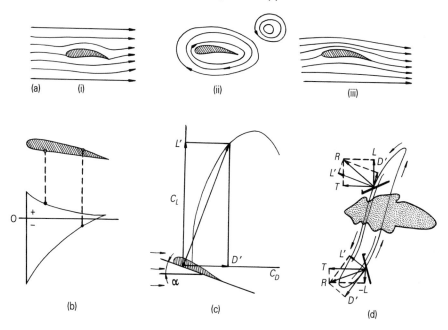

Fig. 3.4 Lift producing mechanism of the flapping wing: a quasi-steady state approach. (a) In a lift-producing mechanism, air (i) combines with air circulating around the wing (ii) to give the airflow pattern in (iii). In (iii), the air flows smoothly off both the upper and the lower surfaces of the wing at the trailing edge. (b) Pressure distribution around a typical wing profile at a small angle of attack. The ambient pressure is taken as zero. (c) The resultant force on a wing is resolved into a lift force normal to the airflow, and a drag force parallel to the airflow. With the change in angle of attack the resultant force vector describes the curve known as Lilienthal's polar. (d) While moving along the trajectory, settled at a given angle relative to the body axis, the wing is affected by the resultant force which, when integrated over the wing cycle, is responsible for the creation of the sustaining force (lift) which keeps the insect aloft and of the thrust which drives it forward. The forces acting on the cross-section of the wing as well as the forces acting on the insect body are shown for the wing trajectory sections (double line) where the wing moves with a speed and an angle of attack which vary only slightly.

interaction of the moving wings with the airflow, the airflow accelerates and is forced down and backwards. The pulses of force received by the insect direct it forwards and upwards. An evaluation of the forces in the pulses of airflow which are thrown off by wing flapping is widely used in investigations of a special mode of flapping flight, hovering. Just as plane wing theory is used to gain an understanding of forward level flight, so propeller theory is used to describe hovering flight. All propeller theories come down to the formation and throwing back of an air jet. In this approach, the interactions

of the wings with the airflow are not taken into consideration and are looked upon as a black box, in the output of which there is an airflow which is accelerated by the working wings (Fig. 3.5). Since the insect 'hangs' stationary in the air at one point so that the aerodynamic force created by wings counterbalances the body mass ($T = \frac{1}{2}\rho A l^2 U^2_S$, where Al^2 is the area of the wing disk), then the flow velocity in the wake of the jet is $U_S = \sqrt{2G/\rho A l^2}$, where G is the body mass.

The amount of energy spent on the creation of an accelerated flow depends on losses due to overcoming the induced and profile drag and

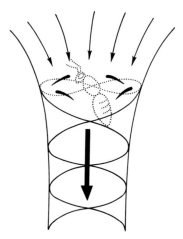

Fig. 3.5 Schematic representation of the flow through the 'wing propeller' of a hovering insect—a momentum jet approach. Due to the acceleration of the trapped air masses, a resultant aerodynamic force is induced which is directed upward and opposite to the main wake flow (shown by a thick arrow). Short curved lines on the wing path correspond to the strongly dorsally and ventrally cambered wing profile as it moves forwards and backwards.

to the jet spinning round. If this energy rate is not taken into account, then the relationship between the flow velocity in the wake (U_S) and the area of the wing disk ($A1^2$) follows from the formulae given above, that is, at a given flapping velocity an insect with long wings, a large stroke angle, or both, can compensate for a larger body mass. Hence insects with high wingbeat frequencies possess relatively short wings, whereas dragonflies achieve the same effect using long wings and low wingbeat frequencies.

As in forward flight, low pressure must be created above the wings and high pressure below them. A jet is produced that is equal in magnitude to the momentum imparted to the air by the reaction of the wing forces. This momentum jet is identical to the vortex wake of the wings. Using this approach, the aerodynamic wake of a hovering insect can be considered as a jet which is spinning round and is being propelled downwards (Fig. 3.5).

However, in a continuous medium like air, the only structure which can retain the momentum passed on to the medium is a vortex ring. Hence vortex rings must be present in the hovering insect's aerodynamic wake. This characteristic of the wake was pointed out by Kokshaysky (1974) and then followed up by Rayner (1979, 1980). According to Rayner, at the end of the active half-cycle of a stroke (the downstroke for the majority of birds in forward and hovering flight, and both half-cycles for hovering insects and hummingbirds—Rayner did not study forward flight in insects), the wings shed the sheet of vorticity which first rolls up into a vortex plait and then into Kelvin's thin vortex ring. The distribution and shape of these rings depend upon the flight mode of the insect or bird.

Let us now consider the flapping wing from the standpoint of the principles laid down in both the approaches discussed above. As has already been mentioned, as soon as a wing begins to move, a starting vortex is formed at its trailing edge. This vortex, through tip vortices, joins the circulatory flow around the wing, forming a ring (Fig. 3.6a). When the moving wing stops suddenly, the boundary layer of the upper wing surface, which moves faster than the boundary layer of the lower surface, flows around the trailing edge and rolls up into a vortex which rotates at the same speed as, but in opposite direction to, the starting vortex. This is called the 'stopping vortex'. In other words, the moving wing carries the vortex ring until it is released when the wing stops moving. If we now imagine that the wing oscillates in the plane perpendicular to the oncoming flow, then the starting and stopping vortices shed from the wing one after another will form a chain of interconnected vortex rings (Fig. 3.6b). The ring formed during the downstroke will carry the force impulse which is directed downwards and backwards, and at the same time the wing will develop lift and thrust because of the difference in pressure between its upper and lower surfaces. Similarly, during the upstroke the wing will develop thrust and negative lift, and the ring formed at this moment will carry the force impulse directed upwards and backwards. Thus perturbations of the air produced by flapping

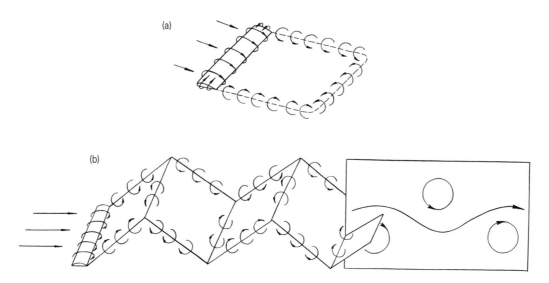

Fig. 3.6 Vortices in the wake of a flapping wing. (a) Simplified vortex distribution of an aerofoil in steady flight. The vortex system consists of a bound vortex (bound circulation) on the wing, two trailing vortices and a transverse starting vortex. The starting vortex is shed whenever the strength of the bound vortex (i.e. the circulation, Γ) changes. (b) The oscillating wing sheds the chain of interconnected vortex rings through the starting-stopping vortices. The section of the vortex wake in the sagittal plane, i.e., a vertical plane passing through the longitudinal body axis, gives its two-dimensional view termed a 'reversed vortex street.'

wings, or rather the aerodynamic wake which the flapping wings leave in the air contains, as if in code, information about the interaction of the wings and the airflows. Wake structure is a key to the understanding of the aerodynamics of flapping flight, and with this key we shall try to open the door leading to the world of the aerodynamics of insect flight.

3.4 Principles of vortex wake formation in flying insects

Before describing the main rules of the interaction of flapping wings with airflow and the organizational principles of the aerodynamic wake, let us consider the dynamics of vortex formation in the wing cycle of the species best known in this regard—the skipper *Thymelicus lineola*. This species is particularly suitable for this purpose because its fore- and hindwings are coupled and move synchronously, its wingbeat frequency is not too high (30 Hz), and the body angle relative to horizontal is zero and the stroke plane angle relative to horizontal is 85°, so that the wings move along a trajectory perpendicular to the longitudinal axis of the body.

To visualize the airflow around an insect in tethered flight, both a cloud of flow visualization particles and smoke were used. Using a flat sheet of illumination to visualize patterns of flow Brodsky and Ivanov (1984) obtained two-dimensional pictures of flow which are suitable for analysis. Flat sections of the flow allowed them to see what was happening at any point in the airspace around the flying insect, and finally to reconstruct the aerodynamic wake in three dimensions. The experiments were all undertaken both in still air and in a wind tunnel.

3.4.1 Vortex formation in the flight of the skipper *Thymelicus lineola*

A downstroke starts with the wings being pulled apart, with the leading edges bent sideways (Fig. 3.7, 1). At this stage the interaction between the wings and the airflow is insignificant, and the air velocity in front of and behind the wings is approximately the same. The rotational velocity of the vortex which forms near the tips of wings is relatively small. Figure 3.7, 2 illustrates the moment when the wings, still pronating, shift forward and slightly downward. The starting vortex increases in size and the velocity of the air motion behind the wings increases sharply. Two small vortices are apparent just behind the abdomen, resulting from stalling of the airflow. Following pronation, the wings begin to flatten at a constant velocity (Fig. 3.7, 3 and 4). During this phase the vortex above the body increases in size noticeably and its rotation velocity increases simultaneously, reaching 1.4 m s^{-1}. The small vortices around the leading edge of the forewing can be seen clearly, as can active streaming of air down the groove in the anal region of the hindwing (Fig. 3.7, 4).

After the phase of uniform and rapid depression the wings brake, and the supination phase begins. This is a critical stage in which all the different processes switch to the opposite direction (Fig. 3.7, 5 and 6). The vortex over the body, still growing, nears the top of the abdomen and actively moves the air right over the insect body. Radial airflow down the wings is clearly noticeable. The following three illustrations (Fig. 3.7, 7–9) show the lifting phase of the supinated wings. The most characteristic traits of the airflow during this phase are the backward movement of the vortex which was previously above the butterfly, and the development and growth of a new vortex under the butterfly's abdomen. Airflow around the wings rolls up into vortices at the leading edges of the wings. The influence of the tip vortex in directing the air jet is striking in the ninth frame (Fig. 3.7, 9).

At the end of the rapid lifting stage, the wings,

not yet at the top of the stroke, pronate slightly and clap their dorsal surfaces together. At the same time, they move forward to reach the maximum position of elevation (Fig. 3.7, 10–12). The vortex under the butterfly's body continues growing, gradually shifting backwards. The final closing-in of the dorsal surfaces of the wings, completing the clap, is followed by a sharp throw of air backwards (Fig. 3.7, 12).

A three-dimensional picture of an aerodynamic wake can be reconstructed by means of a series of flow sections around a flying insects similar to those just discussed. The airflow pictures of all insects with low wingbeat frequencies (up to 50 Hz), except dragonflies, showed similar airflow characteristics in the wake and close up to the insect wings, as follows:

1. In side view (Fig. 3.7, 11), the vortices are arranged in a chequer-board pattern around the central flow, and rotate alternately.
2. In plan view the vortices are located in pairs, rotating in opposite directions on either side of the central flow of the wake.
3. The vortex rotates close to the insect body, over the back during the downstroke (the 'dorsal vortex'), and on the ventral side during the upstroke (the 'ventral vortex') (Fig. 3.8).
4. A small vortex rotates around the leading edge of the wing, in the opposite direction to the dorsal vortex during the downstroke and to the ventral one during the upstroke (see Fig. 3.7, 3 and 9).
5. A vortex rotates near the wing tip both during the downstroke and during the upstroke, directing the airflow down the wing surface and away from the insect's body (see Fig. 3.7, 5).

These discrete vortices are remnants of a vortex ring, transected by a plane of light. The velocity of the airflow in each ring must be increased by the influence of neighbouring vortices. Since the airflow is at its maximum velocity between the insect body and the dorsal vortex during the downstroke, and between the insect body and the ventral vortex during the upstroke, one may assume that vortex rings are formed during both half-strokes, both above and below the body.

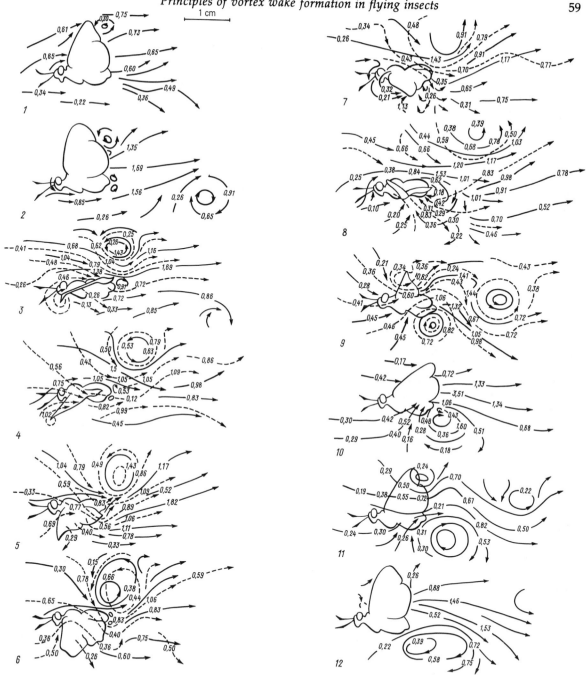

Fig. 3.7 Airflows around the skipper *Thymelicus lineola* during the stroke cycle (1–12). The direction of airflows is indicated by arrows in the vertical plane passing through the longitudinal body axis (continuous line) and through the middle of the left wing (broken line). Figures on the arrows show the airflow velocities in m s^{-1}. Each illustration corresponds to one frame of filming carried out in the conditions described above.

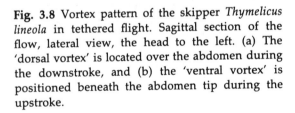

Fig. 3.8 Vortex pattern of the skipper *Thymelicus lineola* in tethered flight. Sagittal section of the flow, lateral view, the head to the left. (a) The 'dorsal vortex' is located over the abdomen during the downstroke, and (b) the 'ventral vortex' is positioned beneath the abdomen tip during the upstroke.

These rings are made up of the leading edge vortex linked by the wing tip vortices to the dorsal vortex during the downstroke, and to the ventral vortex during the upstroke (Fig. 3.9c).

The whole process can be presented dynamically as follows. During the downstroke vortices are generated (the dorsal vortex, and the tip and leading edge vortices), which form a ring above the insect body. At the bottom of the stroke this ring shifts backward, and the circulatory flow that existed around the wings rolls up into a ventral vortex, simultaneously beginning the generation of a new bound circulation around the wings. This upstroke circulation is equal in

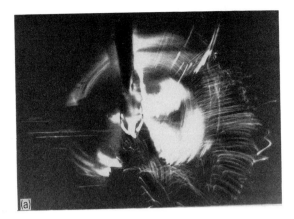

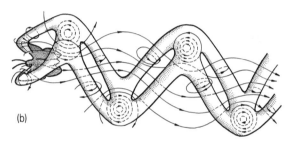

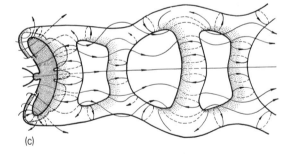

Fig. 3.9 Shape of the aerodynamic wake generated by an insect in forward flapping flight shown in three reciprocally-perpendicular planes. (a) Photograph of a flying skipper, using a vertical plane of light passing through the wing bases. Front view showing the vortex wall of the ring neighbouring the body. (b) Lateral and (c) dorsal views of the vortex wake. The direction of airflow is indicated by arrows.

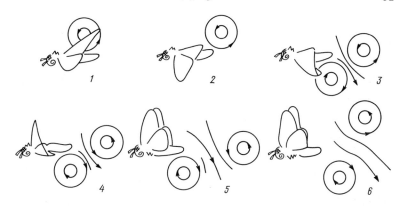

Fig. 3.10 Dynamics of vortex ring in the flap cycle (1–6) of a skipper *Thymelicus lineola*. (See text for explanation.)

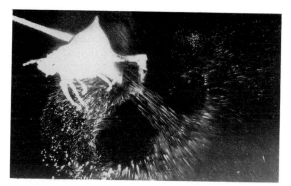

Fig. 3.11 Cross-section of the vortex ring at the middle of the upstroke, showing the jet of air accelerating through the centre of the downstroke vortex ring. (Photo of air currents around the skipper *Thymelicus lineola*, with the head to the left, with the light passing through the sagittal plane).

speed but opposite in direction to the circulation of the ventral vortex. The ventral vortex, being near the insect, enables this new circulation to be generated, since its direction of rotation coincides with that of the starting vortex. As a result, the vortices generated during the upstroke (the ventral vortex, and the tip and leading edge vortices) form a ring below the insect body. At the top of the stroke the circulatory flow around the wings rolls up into the stopping vortex, with which the starting vortex which forms at the beginning of the downstroke becomes connected. Hence, each old vortex ring is linked to each new one via the circulatory flow, so that the wake of a flying insect forms a chain of

coupled rings leaning left and right alternately (Fig. 3.9).

3.4.2 The dynamics of vortices in a flap cycle

The striking resemblance between the wake of a flying insect (Fig. 3.9) and that of a wing oscillating in an airflow (Fig. 3.6) might lead one to believe that the nature of the forces generated by a flapping wing might be explicable using the quasi-steady approach; that is, that the forces result only from pressure differences on the wing at the appropriate phases of the stroke. However, such a conclusion would be over-hasty. Let us consider the vortex dynamics during a flapping cycle.

When the wings are being lowered, the dorsal vortex is situated above the base of the insect abdomen (Fig. 3.10, 1); at the end of downstroke, it shifts toward the tip of the abdomen (Fig. 3.10, 2); and when the wings begin to move upwards, it 'slides off' into a new position (Fig. 3.10, 3). When the wings are supinated at the bottom of the stroke, a stopping vortex whose direction of rotation is opposite to that of the dorsal vortex gradually forms below the abdomen (Fig. 3.10, 3 and 4). The position of the dorsal and ventral vortices remains unchanged during almost the entire time in which the wings are lifting (Fig. 3.10, 3 and 4). The airflow between the two vortices is directed diagonally downwards and backwards (Fig. 3.11). The positions of the vortices change only at the very end of the half-cycle. When wing braking and pronation begin preceding a clap, the vortex ring becomes almost horizontal (Fig. 3.10, 5) because the dorsal vortex shifts backwards. Still clear, though weaker than before, the twisting jet between the vortices becomes almost vertical. Pronation continues even after the wings strike each other's dorsal surfaces. The final stage is rapid opening of the wings. As this begins the ventral vortex moves slightly backwards and the dorsal one moves upwards, forcing the vortex ring to become more vertical (Fig. 3.10, 6).

Thus during the lifting of the wings the vortex ring formed at the downstroke expands and accelerates the airflow downwards and backwards, approximately at an angle of 45° to the horizontal (Fig. 3.11). This twisting jet constitutes a major part of the airflow round the wings during most of the period of wing lifting. The twisting jet reaches its maximum intensity at the middle of the upstroke (Fig. 3.10, 3 and 4) and becomes weaker at the end of wing lifting. In addition, when the wings reach their highest position and begin to pronate, they push the neighbouring ring, and the whole chain of rings with it, backward (Fig. 3.10, 5 and 6). It may be that at this moment the insect receives a small push forward. Whether or not it does, we should be able to be more precise about it in Chapter 9, when we consider the airflow of another species during flight—the peacock butterfly, *Inachis io*.

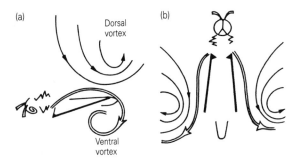

Fig. 3.12 Shedding of vortex rings from the wings (a) at the bottom and (b) at the top of the stroke. (Lateral (a) and dorsal (b) views of airflow around the skipper *Thymelicus lineola*. The vorticity shed from the wing is indicated by double arrows.)

3.4.3 Shedding vortex rings from the wings at the extreme points of a stroke

In spite of the symmetry of the aerodynamic wake of a flying insect, the separation of vortex rings from the wings occurs differently at the upper and the lower points of a stroke. As the wings approach the lowest point of their trajectory, the airflow over their dorsal surfaces deviates towards the tips, passes over the trailing edges of the wings (Fig. 3.12a), and rolls up into two stopping vortices. The plane of rotation of both vortices is approximately perpendicular to the longitudinal axis of the body, and their direction of rotation is downwards and away from the wing base. During supinational wing bending the airflow over the costal edge leaks over on to the ventral surface owing to a 90° change in the plane of rotation of the vortex; the vortices unite and form the ventral vortex, which rotates upwards and backwards (Fig. 3.12a). The ventral vortex closes up the ring shed from the wings. The ventral vortex, as has already been mentioned, acts simultaneously as a stopping and a starting vortex and, as a result, the vortex ring formed during the downstroke joins the new one generated during the upstroke by means of the circulatory flow.

At the highest point of the trajectory the sheet of vorticity shed from the wings rolls up onto a wall of vortex ring (Fig. 3.12b). This vortex ring surrounds the wings through the tip vortices. The pronational opening of wings that follows initiates vortex formation on the dorsal surface of each wing. The starting (dorsal) vortex is formed between the wing surfaces, rapidly increases in size and, together with old tip vortices, completes the ring. At the beginning of a downstroke the dorsal vortex is separated from the wings and shifted backwards. Thus, at the upper point of a stroke the stopping and starting vortices are distinct, lie in mutually perpendicular planes, and are separated by the wings. While at the beginning of the upstroke the ventral vortex generates a new circulation around the wings, the circulation has to be newly and independently created at the beginning of each flap cycle. This fact explains the great significance of the pronational phase in the flight of all insects.

3.4.4 Characteristics of the coupled vortex ring model

The peculiarities of the interaction between flapping wings and the air discussed above suggest that the aerodynamic wake generated by a middle-sized insect with relatively low wingbeat frequency might be modelled as a chain of coupled vortex rings. The main principles underlying the formation of such an aerodynamic wake are that:

(1) the circulatory flow around the wings forms part of the vortex ring;
(2) ring coupling results from periodical changes in the circulation around the wings.

Ring coupling is the most important characteristic of the coupled vortex ring model (Fig. 3.13a). Rings of equal thickness are formed at both the downstroke and the upstroke, and are connected to each other by the starting vortex. The starting vortex that closes up the rings at the lowest point of the stroke results from the circulatory flow shed from the wings, and hence it is also a stopping vortex. A periodical change in direction of circulation around the wings is a necessary condition for ring coupling: at the downstroke the angle of attack is positive, and at the upstroke it is negative (Fig. 3.13a). If, as in the case of some birds, the angle of attack during the upstroke remained positive, then the rings would act as brakes for each other and the chain would not form (Fig. 3.13b).

The ring is rather thick, so that the ratio of the ring diameter[1] to its cross-section is relatively small, and considerably smaller than those of Kelvin's classical rings. The same thick rings were recorded by Kokshaysky (1982) in the flight of a chaffinch. The relative thickness of the rings and their manner of coupling mean that the calculation of forces should be based not on the momentum of individual rings or the sum of all of them, but on the interactions of all the chained rings neighbouring the body.

A sagittal section through the wake gives

[1]Ring diameter is the distance between the centres of the vortices rotating towards each other.

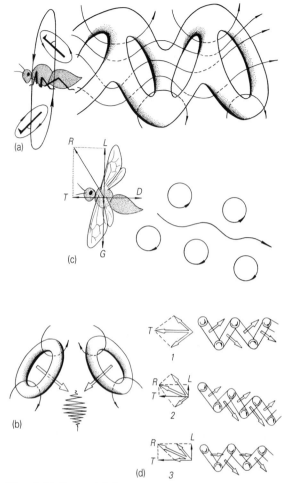

Fig. 3.13 Essential features of the coupled vortex ring model. (a) The aerodynamic wake of the flying insect acquires the shape of a chain of coupled vortex rings if the circulation around the wings changes to the opposite direction at each half-stroke. (b) Without changes in the direction of circulation, the vortex ring coupling could not take place. (c) Reversed vortex street and the forces acting on a flying insect: R is the resultant force; L, lift; T, thrust; D, drag; and G, body weight. (d) Types of reversed vortex streets: 1, straight; 2, slanted; and 3, false straight. If the insect is in steady and level forward flight, the forces acting on it must be in equilibrium: the body and the profile drag, and the insect's weight must be balanced by the reaction of the vortex ring momentum; both ring momentum and its reaction are shown by open arrows.

a two-dimensional longitudinal view, the so-called reversed vortex street. The section consists of two rows of vortices arranged in a chequer-board fashion and rotating alternately (Fig. 3.13c). The vortex street parameters (its width, distance between neighbouring vortices, velocity distribution, etc.) reflect the essential characteristics of the interaction of the wings with the airflow in the creation of aerodynamic forces. Different types of reversed vortex streets are described below.

1. A *straight vortex street* is formed when the ring diameters are equal and symmetrically arranged in the wake relative to the street axis (Fig. 3.13d,1). In addition, the street is oriented perpendicularly to the stroke plane and its axis coincides with the vector of total aerodynamic force acting upon the insect. At the downstroke, lift and thrust are produced. The vortex ring formed at this moment will carry a force impulse directed downwards and backwards. The twisting jet (*a*) created at the downstroke is oriented in the same direction (Fig. 3.14a). During the upstroke, the negatively signed circulation around the wings generates thrust and negative lift; a corresponding ring will carry a force impulse directed upwards and backwards, in addition

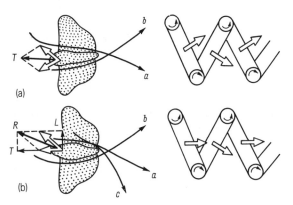

(a)

(b)

Fig. 3.14 Orientation of vortex rings and twisting jet streams (*a–c*) in the wake of a flying insect. Both ring momentum and its reaction are shown by open arrows. (a) A straight vortex street. (b) False straight vortex street.

to the twisting jet (*b*) which is created at this stage (Fig. 3.14a). The resultant of impulses of vortex rings adjacent to the body, and directed backward, in accordance with Newton's third law, defines the creation of thrust (Figs 3.13d, 1 and 3.14a). A straight vortex street has been recorded behind the subimago of the mayfly *Heptagenia sulphurea* (Brodsky 1988). This type of vortex street may also be formed in those insects which fly at a large body angle with their stroke plane 'fixed' relative to the longitudinal axis of the body. In addition, a straight vortex street has been noted in several experiments with a tethered skipper flying in still air.

2. A *slanted vortex street* is formed when the rings are of different sizes. Its axis coincides with the vector of total aerodynamic force, but it is not perpendicular to the stroke plane (Fig. 3.13d, 2). Rings shed by the wings at the end of an aerodynamically active half-stroke have relatively small diameters. For example, if the downstroke is more active than the upstroke (i.e. the wing velocity is higher or the angle of attack is greater), the vortex street is deflected downwards because of the difference in size between the rings leading to some gain in lift but some loss in thrust. A slanted street forms behind a mechanical wing model when the velocities of the up- and downstrokes are different (Polonsky 1950). Its formation has also been recorded in the tethered flight of functionally four-winged insects such as primitive caddisflies, scorpionflies, primitive moths with uncoupled wing pairs, and others.

3. A *false straight vortex street* is observed when a ring with a small diameter, created during the downstroke, expands during wing elevation. This type of street is oriented perpendicular to the stroke plane, and its axis does not coincide with the vector of total aerodynamic force acting on an insect (Fig. 3.13d, 3). In this case, as in the case of the slanted vortex street, the downstroke is more active than the upstroke; however, due to the expansion of a small diameter downstroke ring during the upstroke, the wake itself appears to be a straight street.

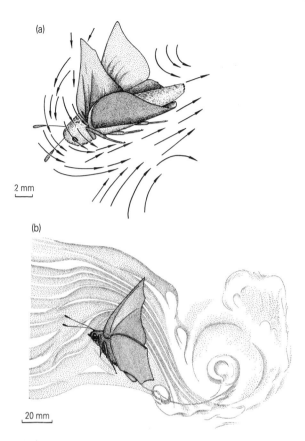

(a)

2 mm

(b)

20 mm

Fig. 3.15 Airflow around a flying insect at the end of the upstroke, side view. Flow lines are shown in the sagittal plane. (a) The skipper *Thymelicus lineola* in tethered flight in still air; (b) The peacock butterfly, *Inachis io* in tethered flight in a wind tunnel with an air speed of 1.0 m s⁻¹.

The widening ring provides acceleration for the twisting jet which is directed obliquely downwards, supposedly compensating for the negative lift created during the upstroke. In summary, the process of force production in one flapping cycle is as follows: lift is created during the downstroke, while thrust is generated throughout the cycle. A third twisting jet (*c*) is added to the two which correspond with those in the straight vortex street (*a* and *b*) (Fig. 3.14b). These conclusions concerning the distribution of forces during the stroke cycle conform with experimentally measured instantaneous force values in the stroke cycle of the desert locust (Cloupeau *et al.* 1979). The false straight street forms during the flight of the majority of insects which have relatively low wingbeat frequencies, as will be shown later.

Fig. 3.15 shows the differences between straight and false straight streets. Both insects, the Essex skipper and the peacock butterfly, were photographed at the same moment of the upstroke, when the wings were approaching the top of the stroke. The Essex skipper is clearly undergoing the effects of negative lift, due to the lack of twisting jet *c* in its wake. When flying in still air, the Essex skipper produces a straight vortex street. In contrast, the peacock butterfly uses a twisting jet *c* to produce substantial lift during the upstroke.

Thus, an explanation of the forces created by flapping wings can not be limited to a quasi-steady aerofoil action. During the upstroke non-steady forces prevail; at that stage of the airflow round the wings, the current created by the previously formed vortex ring dominates. At the upper point of the stroke a kind of 'reactive force' may be added to the non-steady ones. As wingbeat frequency increases, an increase in non-steady effects is to be expected to create the forces necessary for flight.

4

Flight and behaviour

The first winged insects had limited flight abilities, and the range of behaviours which they could perform in the air was not large. However, among the first flying insects there were still some masterly fliers. The dragonflies provide an excellent example of the perfection of ancient flight; they have changed very little from their ancestors, which hunted in the air space above the swampy forests covering the earth about 300 million years ago. During the Carboniferous period primitive cockroaches, mayflies, dragonflies of all sizes, and clumsy Palaeodictyoptera made their way through the air. Swooping and dancing along the edges of lagoons and lapping seas, those early fliers logged more then 200 million years of flying time before the first test flights of the birds, making them airworthy by any standard. Nevertheless, competition with other flying animals—pterodactyls, birds, and bats—proved to be so strong that many groups of insects were forced to limit their claims of supremacy in the air. Some species spend almost all time in the air during adulthood, while others, taking off only for a short period, adopt a relatively immobile way of life the rest of time; the third group have completely lost their ability to fly. Because of this, one of the first steps in the study of the evolution of insect flight needs to be the analysis of flight regimes. In addition, any non-formal classification of flight and of the wing apparatus needs to consider the role that flight plays in the ecology of a given species.

4.1 Flight regimes

The character of an insect's flight depends, on the one hand, upon the structure of its wing apparatus and, on the other hand, on its behaviour. The various components of the general flight pattern of any given group of insects result from the adoption of the behaviour most likely to ensure its continued existence. Insects' ability to fly evolved as an adaptation for capturing prey, or for searching for food or mates. The insect's particular flight regime is a function of the selective value of the flight.

A flight regime can first be understood in terms of the work of the wings, which in turn affects the flight speed, trajectory, ability to undertake different manoeuvres, as well as other features which between them define a broad picture of flight. The sharpest boundary lies between the two most basic regimes of flight—passive flight, in which wingbeats are absent and the insect moves by inertia; and active, i.e. flapping flight.

4.1.1 Passive flight

Both historically as well as in individual cases, passive flight is a continuation of flapping flight; the wings do not create thrust and the insect moves at the expense of energy accumulated earlier, during flapping flight. Gliding in still air entails using the potential energy of a body elevated above the ground to move horizontally. This is achieved by moving down a glide path set at an angle θ to the horizontal (Fig. 4.1). To be able to glide, an insect must first expend energy

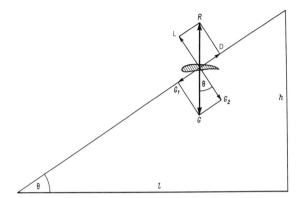

Fig. 4.1 Geometrical and aerodynamic aspects of gliding flight. The insect (a cross-section of the wing in gliding position is shown) moves along the gliding path. The resultant force R is resolved into lift (L) and drag (D) forces. The insect glides along its flight path at a steady gliding velocity, losing height and travelling horizontally. G_1 is the body mass component directed along the gliding path; G_2 is the body mass component normal to it.

gaining height. The following forces act upon an insect during passive flight: 1) gravity; 2) the full aerodynamic force caused by air flowing around a moving body with outspread wings. If the motion is steady, the condition $R = G$ should be satisfied. During gliding flight, the body mass component directed along the path of displacement acts as the motor force (Fig. 4.1). Consequently, only relatively large and heavy insects are capable of passive flight. On average, wing area increases with increase in body mass (Fig. 4.2); however, not all large insects are capable of passive flight. Species whose wings are wide at the base, which provides stability against rolling, are at an advantage in gliding flight.

Passive flight can be subdivided into parachuting, diving, gliding, and soaring according to the displacement trajectory, speed of movement, and the adaptive strategy used. Parachuting, a slow vertical descent, involves the simplest interaction between the wings and the air. It occurs in two unrelated groups: mayflies, and longhorned moths of the genus *Adela* during swarming flights. Swarming males take off vertically, working with

the wings and then, holding them in a somewhat elevated position, descend slowly. Male *Ephemera vulgata* fly up and parachute down in a span of about 1 m. During the descent, the angle between the wing chord and the direction of the oncoming airflow is large (90°), so that the drag causes the

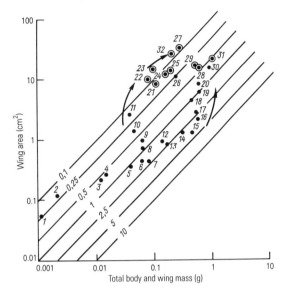

Fig. 4.2 Wing area plotted against flight mass for a number of insects.

1, midge; 2, gnat; 3, blow fly; 4, house fly; 5, sand wasp; 6, unladen honey bee; 7, laden honey bee, 8, carrion fly; 9, hover fly; 10, crane fly; 11, damselfly; 12, bumble bee *Bombus hortorum*; 13, common wasp; 14, bumble bee *Bombus terrestris*; 15, scarab *Amphimallon solstitialis*; 16, carpenter, *Xylocopa violacea*; 17, hornet, *Vespa crabro*; 18, hawk moth; 19, cockchafer; 20, lamellicorn beetle; 21, dragonfly *Calopteryx virgo*; 22, cabbage white (female); 23, cabbage white (male); 24, red admiral; 25, swallowtail butterfly *Papilio machaon*; 26, eggar (male); 27, swallowtail butterfly *Papilio sinon*; 28, dragonfly *Libellula quadrimaculata*; 29, dragonfly *L. depressa*; 30, eggar (female); 31, dragonfly *Aeschna cyanea* (female); 32, monarch butterfly, *Danaus plexippus*. 1–30 after Hertel (1966), 31 after Ryazanova (1965), 32 after Gibo and Pallet (1979). The increase in gliding ability is shown with curved arrows. Data points of insect species which use gliding flight are circled.

wings to work as a parachute and slow down the insect's rate of sinking. The long cerci of mayfly males serve the same purpose, as do the long outstretched antennae of male longhorned moths. During take-off, the mayfly holds its body vertical with its cerci downwards and held together, and when descending the body is held horizontal with the cerci wide apart and with their tips bent upwards.

The principle of gliding flight, which is no different from the gliding flight of an airplane, lies at the basis of the three other types of passive flight. How diving, gliding, and soaring may have arisen from flapping flight during evolution is shown in Fig. 4.3. In all these cases the airflow comes over the wings at a certain angle of attack, while the full aerodynamic force acting upon an insect may be divided into two components: drag and lift (see Fig. 4.1). The lift is directly proportional to the size of the wing area and the square of the gliding speed. Hence two different tactics can be used in gliding flight: diving and gliding. Diving consists of a fast descent at a large angle between flight direction and the horizontal. It is very rarely observed in insects, and only in the Orthoptera and some broad-winged Lepidoptera. Locusts resort to diving to change flight height in a swarm and also to escape danger. A scared grasshopper jumps, spreads its wings, and rises along an inclined trajectory during flapping flight, then it dives with a noticeable loss in altitude, and finally lands. The distance travelled by the blue-winged grasshopper, *Oedipoda coerulescens* when diving from a height

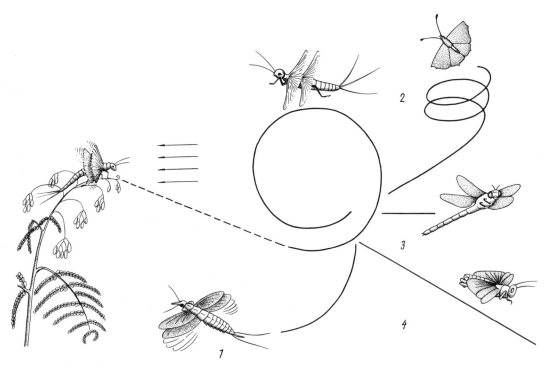

Fig. 4.3 The interrelation between passive and active forms of flight during insect evolution. The circle at the centre symbolizes active (flapping) flight from which different forms of passive flight have evolved in the process of evolution: 1, the peculiar flight of Palaeodictyoptera (very small wingbeat amplitude and frequent using of gliding); 2, soaring; 3, gliding; 4, diving. The initial flight of a 'downward flier' is shown by a broken line (for details see Chapter 5). Arrows indicate the air currents flowing over the head of the descending insect.

of 1 m ranges from 1.5 to 2 m. The same behaviour (but without the jump) has been recorded in *Callimorpha* arctiid moths which live in grass, and in some broad-winged noctuid moths of the *Leucanitis* group. They dive at high speed, with their wings held slightly backwards.

Gliding differs from diving in that it occurs when an insect reaching full speed in a fast flapping flight prolongs its forward movement without wingbeats, thus economizing energy. As in diving, the initial speed is relatively high, but the large wing area plays the main role in the creation of a aerodynamic force sufficient to counterbalance the body mass. The distance travelled during gliding flight is further the smaller the angle between the horizontal and the trajectory of sinking. The tangent of this angle, termed the 'gliding angle', is equal to the ratio of the initial gliding altitude (h) and the distance (l) (see Fig. 4.1). An insect can profitably reduce the angle of gliding and thus increase the distance travelled until contact with the substratum. An insect with a large wing area is at a greater advantage in these conditions. In addition, insects which glide tend to have a low wing loading, i.e. a low ratio of body mass to wing area (Fig. 4.2). Anisopterous Odonata and Rhopalocera typically have the smallest wing loading among all large insects, and form the majority of all insects capable of gliding. These insects are adapted to use ascending air currents for passive flight, i.e. for soaring.

The 'gliding coefficient', or tangent of gliding angle, is proportional to drag and inversely proportional to lift: $\mu = \tan\theta = D/L = h/l$, where D is drag, and L is lift. The higher the lift relative to drag, the longer is the distance travelled before contact with the substratum. Consequently, the simplest way to reduce the angle of gliding is to reduce drag. A considerable proportion of the body and wing drag is made up of the induced drag. Other components of the total drag experienced by a gliding insect are the pressure (form) and friction drag of the wings (conventionally termed profile drag), and that of the body (termed parasite drag). The induced drag is the drag component incurred because lift is being produced and arises from the wing tip vortices. Due to the difference in pressure, the air

flows via the wing tip from the lower surface to the upper one, as a result of which the pressure equilibrates, and lift drops. The tip vortices are more developed on short wide wings than on narrow long ones of equal area. Consequently, the narrower wings have less induced drag.

It turns out that insects which use gliding often, usually have long wings with a sharp tip. Long narrow wings with a high aspect ratio are typical of most species of Heliconiidae and for some Papilionidae (for details see Chapter 9). There are also a number of pierid butterflies with higher aspect ratio for which gliding predominates in flight. The Nymphalidae, Danaidae, Papilionidae, and other butterflies also often resort to gliding flight. According to Magnan (1934), the large fritillary, *Argynnis pandora* can glide for 20 s at a speed of 1–3 m s^{-1}. During gliding its wings are held slightly raised.

Anisopterous Odonata are also good at gliding. Thus, any dragonfly of the genus *Aeschna* is capable of gliding for up to 0.5 min, almost without any loss in altitude. Then, after several flaps of its wings, it begins to glide again. During this time, the forewings are slightly held raised but the hindwings are spread out flat. The hindwings keep the same position as when the dragonfly beats its forewings to withstand strong gusts of an oncoming wind. The hindwings are broadened at their bases and predominate during gliding.

An interesting mode of flight, in which the wings perform extremely rare beats (1–2 strokes per second) with a large amplitude, is normal for calopterygid dragonflies. Such a flight regime is intermediate between flapping and passive flight.

Soaring is passive flight in which the insect makes use of ascending air currents. Irregularities in the landscape or uneven heating of the air cause the ascending flows necessary for soaring. As the wind meets an obstacle it begins to flow round it and upwards. When the ground is unevenly heated, ascending currents of air spread out at the top to form a vortex ring, on the surface of which the animal glides in circles (Fig. 4.4).

As has already been mentioned, only large insects with low wing loading are capable of

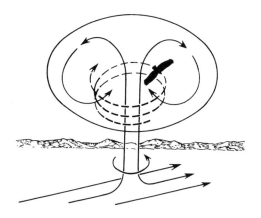

Fig. 4.4 A vortex ring triggered by heated ground, rising as a discrete bubble. The circulation of air up the middle and down the outside of the vortex ring continues for half an hour or more, so that a bird or a butterfly can climb more than 2500 m in such a thermal (after Pennycuick 1972).

soaring and gliding flight. The wings of such insects operate around the Reynolds number at which there is a real danger of a leading edge stall. If the angle of attack exceeds a certain value, termed the critical angle of attack, a leading edge stall takes place. Usually the airflow stalled by the leading edge rises, rolls, and forms a large vortex occupying almost the entire upper surface of the wing. As a consequence of this, drag increases sharply and lift drops. Flight becomes impossible. Hence the wings of insects which are capable of gliding flight must possess special devices which prevent stalling or which minimize its harmful consequences. With this in mind, investigators of insect flight have examined the role of wing pleating in mayflies (Brodsky 1970*a*) and in dragonflies (Newman *et al.* 1977) and also of the scales covering butterfly wings (Nachtigall 1967). Lepidoptera demonstrate a very broad spectrum of adaptations to gliding flight. In the course of evolution they have developed a number of mechanisms which permit gliding at very high values of angle of attack, without the stall which is inevitable in such cases. It is not surprising that such insects actively use gliding and soaring, often including

them into their customary behaviour patterns in the air. Gibo and Pallett (1979) have shown that during its autumn migration the monarch butterfly, *Danaus plexippus* uses both the air currents over the high points and the thermal currents. Using the former, it can glide with motionless wings, flying more than 50 m in a straight line. But moving in circles or in a spiral on ascending currents it can climb to a great altitude.

4.1.2 Active (flapping) flight

However interesting the adaptations for passive flight, flapping flight is nevertheless the main form of locomotion of insects in air. It is used by all winged insects capable of flying.

In general, the forward flapping flight of insects could be described as manoeuvrable, economic, steady, and fast, though the extent of each of these varies across groups. Some insects, such as the migratory locust and large beetles, are capable of rectilinear motion in the air, but in other insects straight sections are rarely more than a few centimetres long. Large insects are able to fly in a straight line for relatively longer periods of time than small ones, and this is true in the Coleoptera more than in other insects; it is to do with stabilization of the trajectory by the outspread elytra. In most species, the flight path is made up of all kinds of turns, short jerks to the side, loops, doubling back, sudden stops, and sharp changes in height. All the manoeuvres are accompanied by changes in height and are usual for forward flight. Because of this, the trajectory of horizontal flight is often wavy.

A striking feature of the forward flight of insects is the pronounced irregularity of its speed. At certain moments an insect accelerates sharply, while at other moments it stops or continues flying at a half or a third of its previous speed. Because of this, the meaning of 'average' speed loses its ethological sense; it is only possible to speak of the range of speeds available for a given species. The average speed ranges between species from several centimetres per second to 15 m s^{-1}. The maximum speed of level forward flight is attained by such insects as anisopterous Odonata, sphinx moths, and brachyceran Diptera.

The speed of insect flight is dependant upon the wind. It is usual to distinguish between absolute speed (air speed), i.e. the speed of flight in calm weather, and relative speed (ground speed) which depends on both air speed and wind direction. Addition of the vectors of the wind speed and the air speed of the insect gives the ground speed. Clearly, in a favourable wind an insect can reach a high speed of locomotion relative to the ground. Thus, with a fair wind the monarch butterfly can fly at more than 14 m s^{-1}.

Manoeuvrability can be defined as the speed at which the direction of motion can be changed. Acceleration during flight is most developed in the Diptera—in hover flies and bee flies. Dragonflies are slightly inferior for this. In some ethological classes of flight, dragonflies of the genus *Libellula* are able to change their speed so quickly that their acceleration is equal to 2.5 g. The fastest acceleration in dragonflies (up to 9 g) has been observed in the family Corduliidae (Ryazanova 1966). The lowest manoeuvring speed among dragonflies is peculiar to members of the family Calopterygidae: a 90° turn lasts 0.5 s, but these turns can follow one another without pause, forming a cascade. Acceleration at the transition from hovering to forward flight in hover flies and in bee flies reaches 175 m s^{-2}, which is up to 18 g. At the transition to upward flight the acceleration is 44 m s^{-2}, or 4.5 g (Termier 1970).

During hovering flight, unlike in forward flight, the insect maintains its position in space, by hanging. 'Hovering' is the widely accepted term for this special form of flapping flight. Hovering is used by:

(1) large insects that feed on nectar, but do not land on the flower because of their heavy bodies;

(2) diurnal predators, which sometimes have to stop in the air to see small prey;

(3) insects which have to stop in the air to search for a mating partner;

(4) insects for which hovering is included in courtship.

Sphinx moths, dragonflies, some bugs and beetles, and all Diptera and Hymenoptera are capable of hovering. The trajectory followed by the wings of a hovering insect may be inclined at various angles to the horizontal. Three types of hovering are distinguished on the basis of this. In the first type, the stroke plane is oriented horizontally, which is achieved by the insect taking an inclined or sometimes even a vertical body position. The wings are moved with a large amplitude and their tips follow a figure-of-eight shaped path (see Fig. 3.5, p.56). At the forward and rearward extremes of the trajectory, the wing rotates rapidly around its longitudinal axis and passes both parts of the trajectory with equal absolute values of angle of attack, and with the vector of full aerodynamic force directed upwards. Because of this feature the movement of insect wings is reminiscent of the rotation of a helicopter screw. The aerodynamics of hovering flight with a horizontal stroke plane was studied by Weis-Fogh (1972, 1973), who named it 'normal' hovering. A variety of insects, such as crane flies, hover flies, bees, sphinx moths, some beetles, and others often use normal hovering in their flight pattern.

In the second type of hovering—with an inclined stroke plane—the body of the insect is oriented horizontally, its stroke plane forms an angle with the horizontal, and the wingbeat amplitude is extremely small (Fig. 4.5). It is believed that only members of two groups of insects hold themselves thus: hover flies of the subfamily Syrphinae and anisopterous Odonata. My observations in the wild suggest that this type of hovering is also used by honey bees and common wasps.

The third type of hovering, with a vertical stroke plane, was described by Ellington (1980) in the cabbage white. In this case the wings move along the trajectory perpendicular to the horizontal: downwards with an angle of attack of about 90°, and upwards with an angle of attack of 0°. A vortex ring is shed from the wings of the butterfly at the lowest point of the stroke. It should be noted that this type of wing motion in the cabbage white was observed by Ellington as it took off (at the first wing stroke); this type of wing functioning is peculiar to the Pieridae and some other broad-winged Lepidoptera during migratory flight, but none of them is capable of real hovering, i.e. prolonged and steady flight in one spot.

4.2 The ethological classification of flight

It is difficult to overestimate the importance of the role of flight in the life of insects. The reasons for flight are various. It is traditional to pick out three types of flight: trivial, swarming, and migratory. The two latter forms have characteristic features that distinguish them from trivial flight.

4.2.1 Trivial flight

A trivial flight is often understood as the local displacement of a single insect searching for food, a mating partner, or an oviposition site, or escaping a potential enemy. Trivial flight can include short flights from flower to flower, as well as displacements of several kilometres, as in the flight of lepidopteran males following the sexual attractants of females. Trivial flight serves various purposes: feeding, patrolling, and mating flights (Fig. 4.6). Their order of occurrence as well as the lifestage at which they occur varies between species.

The adult life of many insects starts with searching flights. In the Pieridae and Nymphalidae, these consist of prolonged non-stop displacements of 5–20 m at an altitude of 0.4 m above the vegetation, with frequent changes in direction. When the swallowtail butterfly *Papilio machaon* searches for feeding sites it flies in circles, returning to the same place several times. Insects can cover considerable distances when

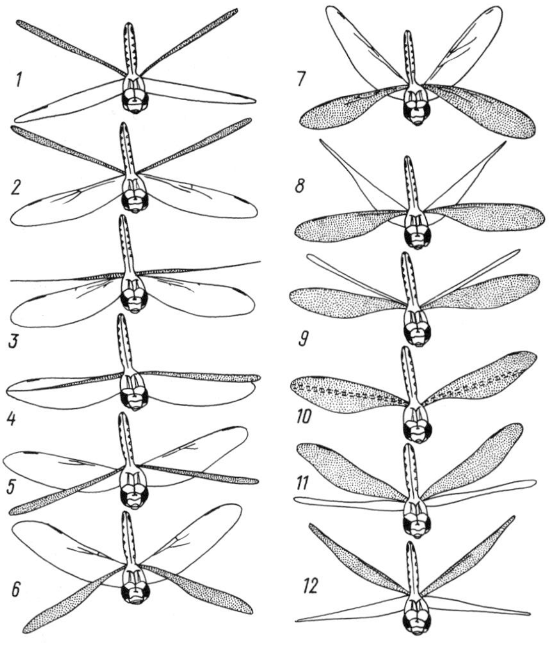

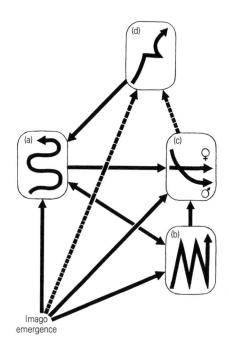

Fig. 4.5 The second type of hovering flight in the dragonfly *Aeschna juncea* (front view). Consecutive film tracings (1–12) from a single wingbeat. Stroke parameters (see list of abbreviations, p.xi): $n = 17$ Hz, $A_{fw} = 60°$, $A_{hw} = 60°$, $b_{fw} = 60°$, $b_{hw} = 60°$, $B = -5°$. The forewings are shaded (after Norberg 1975).

Fig. 4.6 The sequence of ethological categories of flight in the adult life of insects. (a) Feeding flights; (b) patrolling flights; (c) mating flights; (d) migration flights. Searching flights are indicated by continuous arrows and displacement flights by broken arrows.

looking for suitable oviposition sites, feeding sites, or for sexual aggregation. Thus females of the pierid butterfly *Euchloe ausonides* usually lay one egg on a plant, and since plants suitable for these purposes are located far apart, they have to fly up to 1460 m per day (Scott 1975).

Feeding flights proceed differently in plant-feeders and predators. Dragonflies of the family Libellulidae hunt for their prey by laying in wait on a branch or plant sticking out of the water. The appearance of potential prey elicits a fast take-off and an attack, after which the dragonfly returns to the same place. Dragonflies of the family Corduliidae fly rapidly around all the inlets and contours of the banks of water looking for prey. Dragonflies of the family Aeschnidae fly over roads or along the edges of a forest changing height and often gliding. Zygopterous Odonata hunt at the edges of rivers, in grassy vegetation, manoeuvring among the stalks from which they gather small insects. Predacious Diptera such as dance flies, robber flies, and others hunt on the wing. Whites and brushfooted butterflies fly slowly among flowers searching for nectar.

Patrolling flights are peculiar to insects whose males occupy territories which they guard against other males (Odonata, Lepidoptera, and Hymenoptera). Males perform periodical flights over their territories, guarding them against rivals and looking for females. Territorial behaviour is strongly developed in all dragonflies except Aeschnidae. Individual areas are also occupied by some male butterflies, such as blues, brushfooted butterflies, and others. Patrolling flights over plants and attacks on males are also known in hover flies. A visit to the territory by a conspecific male or a female leads either to aggressive behaviour or to courtship.

Mating flights usually precede copulation. A male flies up to a female, hovers above her, makes some jerking flights to the side, flies away, flies up again, etc. Such nuptial dances are characteristic, for instance, of robber flies. During the courtship of the forest fritillary, *Argynnis paphia*, the male flies ahead of then above the flying female, lags behind, then repeats this ritual over and over again (Fig. 4.7). This form of flight provides the female with repeated opportunities to sense the sexual pheromone emanating from the male.

The copulation of some insects (such as mayflies, dragonflies, midges, dance flies, and bee flies) takes place in the air. In butterflies that usually copulate on a substratum, a disturbed couple may take off without becoming unhooked. A female blue butterfly then flies off, dragging the male behind her; the male, having lifted its wings, hangs off the end of her abdomen with its head downwards. Similarly, the females of many tiny midges tow the males behind them, in this case with the males' wings folded. When copulating in flight, the two partners may hold their heads facing opposite sides (caddisflies, nemestrinids), one above another (mayflies), or in tandem. Dragonflies spend a lot of time in tandem flight.

The last type of trivial flight is peculiar to insects which have been disturbed. As a rule, high speed and sharp jerks are utilized in escape flights. Small day-flying butterflies fly 10–15 m then hide themselves in the grass. A disturbed clearwing moth takes off vertically upwards and flies quickly away in straight line. This rocket-like flight is also typical for grasshoppers. At the approach of danger, tiger beetles jump and fly 5–10 m in a straight line close to the ground.

4.2.2 Swarming flight

Swarming flight is characteristic of many insects, especially in populations whose individuals are normally dispersed (Diptera), those populations which have a mosaic (patchy) distribution (Isoptera, Hymenoptera), and also in amphibious

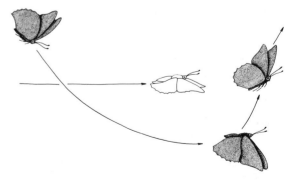

Fig. 4.7 Courtship of the forest fritillary, *Argynnis paphia*. The male is shaded.

insects (Ephemeroptera, Plecoptera, Trichoptera, and Diptera). In these groups, males gather in a swarm which attracts females. Swarming behaviour, in which a group of insects remains stationary flying over one spot, may be regarded as a special instance of insects showing a common reaction to a feature of the environment. The stationarity of swarming insects is achieved by the coordination of flight by many individuals owing to their visual reaction to a particular feature in the environment known as the 'marker'.

Mayflies and longhorned moths of the Adelidae family have the most primitive swarming flight. They maintain a stationary position in space by alternating active flights upward and passive descents. The next step in swarming behavioural complexity is flying in a zigzag trajectory without turning the longitudinal axis of the body in the direction of flight. Some caddisflies which move in such zigzags quickly become displaced from one side of a swarm to another. Other very manoeuvrable insects (some Trichoptera and all Diptera) fly from the periphery to the centre of a swarm and back along the trajectory in the shape of a loop or a horizontal figure of eight. In this case, the longitudinal axis of the body is oriented in the direction of motion. A tangled trajectory of swarming motion, including fast jerks to the sides on the way from the top of the swarm down and towards the periphery, then upwards again through the centre, is peculiar to some Diptera and Hymenoptera; here hovering is often used.

4.2.3 Migratory flight

The distances covered in trivial flight are not large. Thus, male scale insects may fly up to 200 m with the wind, and half that against it. Large insects may fly further on leaving their site of emergence: capsid bugs and checkered beetles travel up to 2 km. The flight of all insects depends upon the wind. An insect's flight activity is suppressed when the speed of the wind exceeds its flight speed. In this case the insect cannot withstand the effect of wind drift and therefore often does not take off. The maximum wind speed at which male scale insects fly is about 0.5 m s^{-1}; ermine moths fly when the

wind speed does not exceed 2.5 m s^{-1}. When the wind is not strong and the insect is in the air, it very often (always when swarming) turns into the wind, making headway against it.

Most adult insects have periods when flight activity begins to dominate their behaviour. It is then that migrations take place. Migration, according to the definition of Johnson (1969), is steady flight uninterrupted by mating and characterized by a certain trend. Taylor (1974) defined the 'boundary layer' as the layer of air above the ground surface in which the insects can control the flight direction and in which their speed exceeds the wind velocity. Once out of the boundary layer, the insect gets into air currents which may carry it for dozens or even hundreds of kilometres. One may therefore distinguish two principal forms of migration: active and passive, i.e. inside the boundary layer in which the insects can control the speed and direction of their flight, and by means of air currents.

Active migration often starts with a short displacement flight which turns into a migratory flight (Fig. 4.6). Displacement flight is typical of aquatic insects (water striders, whirligig beetles, and others) whose adults leave the water pools where they lived as larvae. Other insects, which have a mosaic distribution in the biotope, namely longhorned beetles, bark beetles and others, also make displacement flights. A change of biotope is accomplished by the migration. Thus, females of the cockchafer, *Melolontha melolontha* fly from their site of emergence in fields to the places where they feed and then return to the fields again for egg-laying.

Active migrations are characteristic for strongly flying species. They sometimes undertake long distance transmigrations. During migration they keep close to the ground surface where the wind speed is low. Thus, moths of the genus *Urania* fly at a height of 1–4 m with a speed of 5 m s^{-1}, sometimes making spurts of up to 8 m s^{-1} (Griffith 1972). According to Gatter (1981), hover flies and pierid and nymphalid butterflies undertake seasonal migrations independent of wind direction. The most famous New World pierid migrant is *Phoebis sennae*, whose mass migrations over the USA have been recorded repeatedly.

A curious form of active migration is seen in the upstream flight of amphibious insects such as mayflies, caddisflies, and probably stoneflies over wide clay-bottomed rivers, to compensate for drifting of eggs and larvae caused by water currents. Such flights are taken by gravid females which fly upstream to oviposit. What is unusual about this compensating flight compared with other active migrations is that the insects fly not only against the wind but also in the opposite direction to the water current. How these two not always favourably combined factors, wind and water current, go together, is shown in Fig. 4.8.

Passive migration starts with an upward flight beyond the boundary layer. In many insects, e.g. tussock moths, such flights are performed in regions where the population density is high. The height at which the boll weevil, *Anthonomus grandis* flies over a cotton field is different in spring and autumn. In spring the beetles fly low, at a height of 1 m, but in autumn they climb as high as 120 m, thus promoting passive migration.

Insects migrate for great distances on air currents. The insect creates lift by flapping its wings to counterbalance its body weight, which

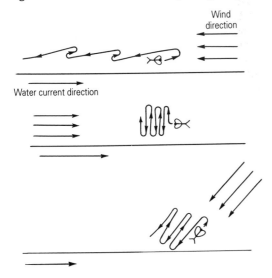

Fig. 4.8 The compensatory flight path of a mayfly female flying up river with different wind directions (after Russev 1973).

increases the distance travelled on the wind by such 'weightless' insects. The aerial plankton is largely composed of small and minute insects: thrips, aphids, mosquitoes, flies, bugs, moths, and many others. During flight aphids, like many other passive migrants, are subject to considerable vertical displacements, which affect the direction and distance of their horizontal movement.

The regularity of passive migrations has been observed in delphacid planthoppers which are carried to Japan on the monsoon winds from the tropics. The bug *Oncopeltus fasciatus*, which is unable to survive northern winters, annually repopulates the moderate latitudes by means of long migrations at great heights on favourable winds.

The longest distance migrations are accomplished by members of three orders: the Odonata, Orthoptera, and Lepidoptera. All of these are capable of travelling great distances using a combination of active and passive flight. Long flights of anisopterous Odonata are well known and take place over dry land and stretches of water. These dragonflies often travel in swarms. The same mode of migration is customary in the Orthoptera. Swarms of flying locusts are carried by air currents, but inside the swarms themselves the insects are flying continuously, moving around within it. There are many locust species, but the best known is *Schistocerca gregaria* which extends across northern and central Africa to the Middle East, Arabia, and India. Swarms of *Schistocerca* often occupy an area of over 10 km², and large swarms may extend over 250 km². This involves vast numbers of insects; one swarm covering an area of 20 km² was estimated to contain up to 1 000 000 000 individuals. The speed of locomotion of a swarm, depending on the wind velocity, can reach 100 km day^{-1}. The migratory instinct occurs not only in locusts, but also in members of other families of the order: Gryllotalpidae, Pyrgomorphidae, and Gryllidae. A swarm of crickets (*Gryllus bimaculatus*) once landed on a ship 933 km away from dry land, having been in the air for about 50 hours (Ragge 1972).

There are especially many migrants among the Lepidoptera. Noctuid moths, leafroller moths, nymphalid, pierid, danaid, and uraniid butterflies migrate most often. A study of migration of the noctuid moth *Spodoptera exempta* using radar (Riley *et al.* 1983) showed that continuous flight of insects at the speed of 1.4 m s^{-1} lasted the whole night. These moths can travel 3500 km in 10 nights.

Among the most famous insect migrants is the monarch butterfly. At the end of summer and at the beginning of autumn it leaves northern USA and Canada for the southern United States, Mexico, Cuba, and the Bahamas, a distance of 4000 km. During their migration the butterflies actively use favourable winds, avoid contrary winds and combine active flight with long glides and soaring. Gibo and Pallett (1979) describe the tactics which the butterflies use in the following way: 'Some butterflies which were flying in a straight course approximately 1–2 m above the ground were observed suddenly to initiate a powerful climb. The butterflies climbed more or less vertically, drifted with the wind, and often flew in an ascending spiral pattern. Upon reaching an altitude of 10–15 m they would stop flapping and began to soar in circles.' Climbing thus, the butterflies were able to find a suitable air current to leave the thermal flow and flew south with the wind. In spring the butterflies migrate back to the north.

4.2.4 Mass flight

The appearance of a large number of flying insects, so common in nature, is usually associated either with swarming or with migration behaviour. The synchronous flight and swarming of a large number of insects is characteristic of termites, march flies, and midges, but first and foremost of mayflies. The mass flight of mayflies attracted attention long ago by the scale of the phenomenon. It produced an indelible impression on everybody who witnessed it.

Insect migrations frequently occur on a grand scale too. Swarms of migrating locusts can be large enough to cover the sun, as may clouds of flying ladybird beetles, etc. Migrating *Libellula quadrimaculata* drogonflies frequently form large swarms. Calculation of the number of dragonflies migrating in the south of France gave a figure of about 400 thousand individuals per square kilometre (Dumont 1964), but their total quantity in Poland was evaluated as 42 thousand dragonflies (Łabedski 1982). The number of individuals in a migrating swarm of the nettle-tree butterfly *Libytheana bachmanii* at peak flight activity reached 2 million insects per hour km^{-1} (Helfer 1972). Knocked down by cars and unable to fly any further, the butterflies moved on to the ground, precisely maintaining their direction of migration. A similar effect is produced by the mass appearance of other insects, such as thrips or aphids. Winged aphids took over the beaches of eastern England in 1979 and landed on the roads in such numbers that they covered the windows of cars, preventing them from being driven.

PART II

The evolution of insect flight

We, who crawled on the ground yesterday, today have risen high into the sky like a swarm of midges, and can spit upon the hat of our enemy from a height of five hundred meters.

Taffy, *And so it has become*

5

The origins of flight and wings in insects

How did insects first evolve those unique structures, the wings, and evolve them so early in geologic time? Here we enter another area of controversy, and one that is not likely to be resolved right way.

S.Dalton, *Borne on the wind*

No structure in the Arthropoda, an extensive group of animals, has given rise to such a variety of hypotheses about their origin as have insect wings. Interest in the more than 150-year-old theories of insect flight has not faded, and is impelled by the importance of the colonization of the air for the evolution of insects. Without understanding the origin of flight we cannot establish the relationships between insects and other groups of Arthropoda nor follow the path of evolution which insects have followed since they acquired wings.

Authors have used a number of methods to solve this problem. One approach is the comparative study of structure of the wing apparatus of unspecialized insects, and investigation of the ontogeny of the wings, skeleton, and musculature, recently added to by an analysis of the development of neuro-muscular relationships in the wing apparatus of relatively primitive insects. However, the study of fossil records still dominates (see the works of Rasnitsyn (1976, 1981), Kukalová-Peck (1978, 1983), and Wootton (1981)).

The traditional palaeontological approach to the understanding of insect flight and wing origin has three important features. First, insects had flown long before the time when the first known winged insects appeared (the Namurian, the first age of the upper Carboniferous). Assuming that the ability to fly arose somewhere between the Devonian and the Carboniferous, 20 million years of the evolutionary development of winged insects are shrouded in mystery. Second, the oldest winged insect fossils known have been found in areas which were tropical marshland, where the conditions for the preservation of fossils were most favourable. Information about other organisms which lived in different environments is very limited. Third, geological records generally comprise abundant species which have a high probability of fossilization. In general, such abundant forms have deviated from the generalized state. Adaptations to specific conditions tend to be accompanied by an outbreak of speciation, whereas less specialized species which retain their generalized features to a great extent are not numerous and the probability of their being discovered is low. An analogous picture is observed for recent insects. Relatively generalized groups such as alderflies, scorpionflies, and others are represented by a small number of species relative to related groups which have none the less colonized some peculiar ecological niches: Coleoptera, Diptera, Lepidoptera, and others.

5.1 Before wings originated

The early stages of the evolution of winged insects have been subject of considerable speculation, resulting in a great number of hypotheses which attempt to elucidate the origin of wings and flight. These hypotheses may be classified according to the organ presumed to be the wing precursor. The most widely accepted are the tracheal, paranotal, and stigmal theories. The proposal by Gegenbaur, at the end of the nineteenth century, that wings originated from the abdominal tracheal gills of mayfly nymphs has been discussed repeatedly in the literature. Various authors have tried to confirm or reject this idea, debating not only the idea that wings originated from gills, but also whether the gills of mayfly nymphs and the wings of insects are homologous. In mayfly nymphs, 'tracheal gills' is the commonly accepted term for the paired appendages on abdominal segments I–VII (in some fossil forms I–IX) which are homologous in nymphs of all mayflies but which may vary fundamentally in function. In many mayfly nymphs these appendages serve to create of water currents round the nymphal body.

Kluge (1989) studied the structure of the tracheal gills (which he called 'tergaliae') of mayfly nymphs and showed that regardless of their specialization they are always articulated with the lateral parts of the tergum: their sites of attachment are homologous in all Ephemeroptera. Moreover, all mayfly nymphs, whether they move their tergaliae actively or not, have muscles that run obliquely forward from the base of the tergalia to the ventral wall of the segment. In literature these are designated 'gill' or 'branchial' muscles. These muscles are the most lateral of all abdominal muscles; the dorsoventral muscles, which point completely vertically, lie medially from them, and in the medial part of the abdomen there are powerful longitudinal muscles which run straight along the body, some of them crossing it obliquely (Fig. 5.1).

It is not possible to compare the abdominal and thoracic segments muscle by muscle because of the remarkable structural differences between the segments. However, the medial longitudinal dorsal and ventral muscles of the abdomen (*dlm* and *vlm* in Fig. 5.1b) are undoubtedly homologous with the dorsal and ventral muscles of the thorax (*t12*, *t14*, *s11–20*, Fig. 5.1a), and the the more lateral dorsoventral musculature (*dvm 1–3*, Fig. 5.1b) is homologous with that of the thorax (*t–p5*, *6*, Fig. 5.1a). Thus, the position of the tergalia muscles (*km* in Fig. 5.1b) is quite similar to those of the direct wing muscles

which originate from the subalar and basalar sclerites (*p3*, *p–s12*, *t–s4*, *5*, *t–p16*, *19*, and *t–cx8* Fig. 5.1a).

The data cited above set out to prove that the paired tracheal gills of mayfly nymphs and insect wings are homologous organs. Among contemporary authors the idea of homology between mayfly nymph gills and wings was upheld by Kukalová-Peck (1978, 1983), who considered

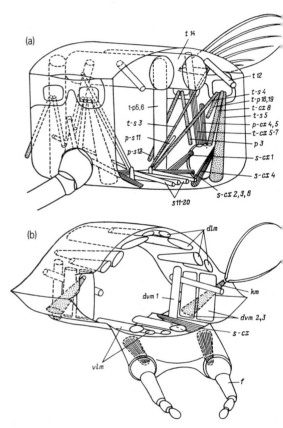

Fig. 5.1 Comparison between the musculature of a wing-bearing and a tergalia-bearing segment. Leg muscles are striped, and the direct wing and tergalia muscles are dotted (after Kluge 1989). (a) Simplified scheme of a wing-bearing segment in insects. (b) Generalized scheme of an abdominal segment in the mayfly. True mayflies differ from this representation in that the tergaliae are only present in nymphs and forceps (f) are only present in the imago.

that both structures originated from movable appendages of the most proximal segment of the leg, the epicoxa. We may then infer that the ancestors of the Pterygota had paired movably articulated appendages—prototergaliae—which originated on the sides of the terga. Their function, shape, and size are unknown because no fossils of them remain. The prototergaliae were presumably flat with two stiff ribs along their edges, and were activated by the muscles projecting from their base to the ventral wall of the body. Authors of the various theories of wing origin usually try to explain the development of wing precursors logically, and to prove that the ancient insects needed these organs. But this conception seems to be erroneous. If the prototergaliae were necessary for an important function, they would have been kept or, having evolved into wings, should have been substituted for by some analogous structure which took over their function. However, if we reject the tracheal hypothesis of wing origin and assume that the primary function of the wing precursors (prototergaliae) was not that of tracheal gills, we have to accept that in all known fossil and present-day insects, wingless, winged, and secondarily wingless, there are no organs which do serve the function of the prototergaliae. All this leads to the conclusion that their function must have been quite specific, and unnecessary for most insects. This is why the function cannot be reconstructed by logical reasoning; among all the possible functions of prototergaliae we suggest that the most probable are unexpected ones such as signalling or for holding eggs on the back.

It is quite probable that prototergaliae were present on all body segments except the first and the last, i.e. from segment II of the thorax to segment IX of the abdomen. In this case, wings homologies have never existed on the prothorax of insects, so that the paranotal lobes of the Palaeodictyoptera, in the nymphs of the mayfly *Ecdyonurus*, in Tingid bugs and many other insects, are of secondary origin.

The prototergaliae, located on the leg-bearing segments (i.e. on the meso- and metathorax) were very powerfully developed and later on evolved into flight organs; as well as two stiff ribs, they gained accessory venation. By that time the primary function of the prototergaliae, which is unknown to us, was lost and the prototergaliae of abdominal segments were reduced. They were retained only in the group of insects which became aquatic in the nymphal stage, and later gave rise to the Ephemeroptera. In these aquatic nymphs the abdominal prototergaliae remained because they started to perform another function, creating a current of water around the nymphal body, and so evolved into tracheal gills. These nymph tergaliae, as well as Pterygote wings, are evolutionarily stable: they are present in all members of the Ephemeroptera, even though in many cases they have lost their primary function of creating water currents and sometimes perform other functions, sometimes most unexpected, or even remain functionless.

5.2 From unknown ancestor to ancient winged insect

Nowadays flight data and the principles of generating of aerodynamic forces are used more and more often to test hypotheses about wing origins (Kukalová-Peck 1978, 1983; Wootton 1976, 1981, and others), but sometimes the lack of reliable information in this field makes it difficult to distinguish possible from impossible mechanisms. We can judge with some confidence which movements of the presumed wing precursors would be favoured by natural selection in the first place, if we accept the principles underlying the generation of aerodynamic forces in flapping flight and the organization and mechanism of the wing apparatus in primitive insects. Previous chapters of this book have been concerned with an analysis of the work of the insect wing apparatus. This data should enable us to find the paths most likely to have been favoured by natural selection both before flapping flight originated and during the next stages of its development.

5.2.1 Gliding or flapping? Flight mechanics and their evolution

No matter what the nature and primary function of the morphological wing precursors were, at a given moment they began to be used for aerial locomotion in the ancestors of winged insects. As has been suggested by the proponents of the paranotal hypothesis (Forbes 1943, Zalessky 1949, and Rasnitsyn 1976, among others), the wing precursors may have been used as immobile or slightly movable structures to provide lift when gliding off plants. Various forms of passive flight were discussed in Chapter 4. Gliding and soaring are usual for large insects and require special adaptations, changes in the structure of the wings and axillary apparatus. Such adaptations exist in strongly flying insects: anisopterous Odonata, diurnal Lepidoptera, some Neuroptera, among others. The simplest passive flight mode is diving, for which the main specialization is broadening of the wing bases, which enhances the stability of an insect against rolling, in which the point at which aerodynamic force is exerted is close to the longitudinal body axis. It follows from this that selection of individuals with gliding surfaces may lead to further broadening of short wing pads, but will not cause their lengthening. This is why gliding mammals (e.g. flying squirrels) and reptiles (e.g. flying lizards) have the adaptation of broadened gliding surfaces but not lengthened ones. A diving trajectory is relatively short and straight; a large angle to the horizontal must be associated with a flight speed greater than in gliding.

The initial speed needed for diving may be achieved either by jumping or by using horizontal air currents. The use of jumping is unlikely since the ancestors of winged insects did not possess jumping legs, while use of horizontal air currents would result in accidental drifting of insects with short proto-wings into the air. Further selection of individuals with short, broad thoracic outgrowths used in diving would lead either to further broadening or to enlargement of various body appendages, such as cerci, antennae, or legs, thus increasing the probability of drifting on air currents. None of this would lead to true flapping flight.

Studies of the fossil records of Carboniferous insects show that palaeozoic nymphs of primitive Neoptera and of all Palaeoptera have articulated wing cases. All recent nymphs differ in having 'fixed' wing cases, secondarily fused with terga for streamlining and protection. This clearly shows that the primitive nymphal wings were first articulated and only later became immobile or 'fixed' (for details see Kukalová-Peck 1987). Echoes of primarily movable proto-wings may be found in present-day insects. Thus, when fifth instar *Locusta migratoria* nymphs are in the flight position, complex motor activity of the future wing muscles with a contraction frequency of up to 20 Hz is nearly always observed (Kutsch 1974) in the grasshopper *Chorthippus longicornis* which is short-winged and incapable of flight, jumping is accompanied by rapid vibration of the wings (Voisin 1976). It is assumed that these wing movements enable the grasshopper to maintain the right body position for landing. Homologous interneurons generating flight activity are found not only in thoracic ganglia, but also in those of the first three abdominal segments (Robertson et al. 1982). Since the neural tissue of recent insects is extremely conservative, this homology suggests that the proto-wings continued as a series of winglet appendages on the abdomen and that proto-wings, as well as their abdominal equivalents, must have originated as mobile structures. Furthermore, wing vibration is common in many insects during mating periods and also during warming up before flight. In neither cases does the insect take off because it holds on to the substratum, and lacks coordination in the contraction of the different groups of muscles necessary for flight.

There is a leaf beetle *Zygogramma suturalis* which was brought from Canada to Russia to combat the weed *Ambrosia artemisiifolia*, which had well developed wings, but did not use them primarily for flying. At the third generation, the beetles started to exhibit a tendency to fly. They climbed up plants, spread their elytra and flapped their wings, but did not fly forward and fell down. Females of the moth *Bombyx mori* work their short wing pads rapidly when running hard; the intensity of these vibrations grows as they reach the edge of a surface. The tarsal reflex is one of the strongest reflexes of winged insects: loss of tarsal contact with the substratum initiates wing motion. All these phenomena can only be understood if one assumes that one of the functions of the proto-wings, which were capable of rapid vibration, was the provision of some rhythmical behavioural act. But since the central mechanism for this had already been developed, there are no principal objections to a change in its motor program under the influence of natural selection. It should be added that the most favoured hypothesis for the origin of wing flapping in birds from gliding is currently being revised (Caple *et al.* 1983).

Many authors use mechanical models which imitate features of the organization of the ancestors of winged insects to test their ideas about the origins of flapping flight. For instance, Flower (1964) studied the selective advantages of forms with paranotal lobes from an aerodynamic viewpoint, and concluded that insects about 10 mm long with short legs may have had a selective advantage at the gliding stage, with increased development of the paranotal lobes. Such a selective route must have resulted in the development of forms with broad wings which flapped at low wingbeat frequencies. It is also apparent that the selection of forms with short proto-wings which were capable of rapid vibration, should favour the development of relatively long narrow wings. To characterize the flight of a proto-pterygote model which flapped its wings at various frequencies, we calculated the aerodynamic coefficients of forces on the wing with a stroke angle of 83°, the stroke plane vertical, and the wing tip following a figure-of-eight shaped path with isometric loops. The upper and lower convergence angles (those between the planes of the wings when elevated or depressed) are both 97°. The calculations cited below were carried out for a lengthened wing (with an aspect ratio of 6), whose prototype was the forewing of the locust, in forward level flight ($V_\infty = 3.6$ m s^{-1}). The wing moved along its trajectory without accelerating, and the geometrical angle of attack, which was positive throughout the downstroke and negative throughout the upstroke was of equal absolute values in both half-cycles of the stroke.

The values of unsteady aerodynamic coefficients are shown in Fig. 5.2. From comparison of the plots it is evident that at a wingbeat frequency of 24 Hz (Fig. 5.2a) the total lift for the entire cycle is about zero, and thrust retains its maximum value at the middle of both half-cycles; the peak of the upstroke thrust is shifted a little to the end of the stroke. In general, the wing produces lift and thrust during the downstroke, and thrust and negative lift during the upstroke. In other words, from an aerodynamic viewpoint the work of the wings of such an insect is identical to the work of wing apparatus with an aerodynamically symmetrical stroke. We know that in the case of stroke aerodynamic symmetry, i.e. where there is aerodynamic equivalence in

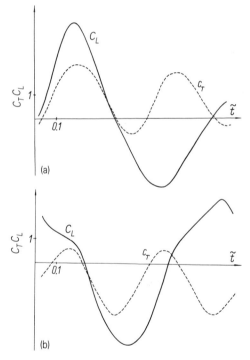

Fig. 5.2 Dynamics of lift and thrust coefficients for an elongate wing performing flaps at different frequencies with (a) $n = 24$ Hz and (b) $n = 14$ Hz. The time axis (abscissa) is averaged over the stroke period. The wing movement starts from the downstroke and continues for a little more than one complete stroke cycle.

both half-cycles of the stroke, the wings of flying insects produce a wake which consists of coupled vortex rings, whose sagittal cross-section forms a straight vortex street.

With a decrease in wingbeat frequency from 24 to 14 Hz (Fig. 5.2b), the dynamics of the aerodynamic coefficients changes remarkably during the stroke cycle. The lift drops abruptly at the middle of the downstroke, and then continues to drop up to the end of half-cycle. The total lift of the stroke cycle is negative. Thrust drops at the end of both half-cycles. The data obtained shows that for the wing of the aspect ratio mentioned, the optimal wingbeat frequency averages 20–30 Hz; with a decrease in wingbeat frequency, the wing is unable to produce positive aerodynamic forces, and flight becomes impossible. The same wingbeat frequency (20–30 Hz) is characteristic of most relatively primitive insects, and may be explained by the physiological peculiarities of tubular wing muscles. Only broad-winged insects, for instance some diurnal butterflies, in which both structure and functioning of flight muscles have undergone considerable changes (George and Bhakthan 1960), flap their wings at a lower wingbeat frequency. Since the wings of the first winged insects were powered by a typically tubular musculature, they had to flap their wings at a frequency of 20–30 Hz. Only wings which are relatively long and narrow could be effective at this frequency, although it is apparent that the aspect ratio of the wings of the first winged insects was not as great as that of the locust.

As well as the obstacles hampering the lengthening of the short gliding surfaces mentioned above, there is one more circumstance that makes movable proto-wings capable of rapid strokes preferable to broad immobile lobes—the relative simplicity of an origin of flapping flight based on the vibration of short movable wing pads. However, once evolved, movable proto-wings inevitably lead to the progressive lengthening of the wings.

5.2.2 Mastering the airspace

In Chapter 3 it was shown that an insect wing flapping at a relatively low frequency produces a system of coupled vortex rings, a sagittal section through which forms a reversed vortex street. The same wake is observed behind the waving caudal fin of a fish or a vibrating mechanical wing model, when a propulsive force is generated. Thus the creation of thrust is always accompanied by a reversed vortex street. Studies of various periodic movements including progressive and rotational oscillations show that bodies which only perform progressive oscillations do not form a reversed vortex street. If the wing rotates relative to the midline of the chord, the vortices in the wake are grouped in pairs and rotate such that they meet in the middle (Hertel 1966); in this case no propulsive force is created. When the wingstrokes are a combination of rotational and progressive oscillations, the thrust increases with the increase in phase angle between them up to a maximum of 90° (Gorelov 1976, Grebeshov and Sagoyan 1976). However, if the wind is head-on to the oscillating body, a propulsive force may only be created by rotational or progressive oscillations. This background information can helpful in discovering the conditions necessary for the creation of aerodynamic forces by the ancestors of winged insects.

What was the oscillational pattern of the short wing-like lobes of the winged insects' ancestors? Study of the structure and functioning of the wing apparatus of relatively primitive insects can shed some light on this question. The main features of the organization of the wing apparatus of the ancestors of the winged insects are as follows:

(1) there are two pairs of short homomorphic symmetrical wing-like lobes;
(2) the stroke plane is vertical to the longitudinal body axis;
(3) the stroke amplitude is relatively large (80–90°);
(4) convergence angles are large and equal at both the upper and lower points of the stroke;
(5) the wingbeat frequency is about 20–30 Hz, i.e. it is equivalent to the physiological capabilities of the tubular type of leg musculature;
(6) the aerodynamic angle of attack is large (about 60°) and is similar during both the up- and downstrokes;

(7) the base of the wing-like lobe is broad, only slightly differentiated, and only capable of providing up and down movements;

(8) movements of the wing-like lobe are powered by the leg musculature, predominantly the pleurocoxal and tergocoxal muscles.

The two most important features of the primitive wing apparatus are:

(1) the full aerodynamic equivalence (symmetry) of the up- and downstrokes, and

(2) the inability of the wing-like lobe to rotate on its longitudinal axis.

Let us start with the first point. That the wing apparatus of the first winged insects was aerodynamically symmetrical is borne out by a number of factors connected with the functioning of the wing apparatus in primitive insects:

(1) the dorsal and ventral surfaces of the wing are morphologically symmetrical (unlike, for instance, in birds);

(2) the wing motion in flight is restricted to up and down movements;

(3) the aerodynamic forces produced in the up- and downstrokes are symmetrical, which results in pitching movements of the insect body as a result of the wingstrokes;

(4) the wing flapping velocity is equal in both up- and downstrokes;

(5) the wing profile is convex throughout the downstroke and concave throughout the upstroke.

One further important feature is that the wings of the first winged insects performed simple flapping motions in a vertical plane, but were incapable of strong rotation about their longitudinal axes. This conclusion is supported by two lines of evidence:

1. The course of evolution of winged insects is continuing in the direction of a gradual decrease in stroke amplitude and an increase in angle of wing rotation at the end of the up- and downstrokes during pronation and supination. Because of their inability to rotate the wing about its longitudinal axis, relatively primitive

insects are incapable of hovering. It is beyond any doubt that in the course of evolution progressive oscillations of the wings were replaced by rotational ones.

2. The hinge which moves the wing in the vertical plane (the horizontal hinge) is located between the wing and the notal margin in all winged insects, i.e. on the edge of a recently developed evolutionary structure, the wing. In all recent insects (excluding dragonflies) this hinge possesses one degree of freedom, its mechanism being like that of a door hinge. By contrast, the hinge which rotates the wing about its longitudinal axis (the torsional hinge) projects into the wing lobe and is characterized by a great variety of positions and structures; it may be made up of different elements.

Hence the ancestors of winged insects, whose wing lobes were incapable of rotational movements about their longitudinal axes, must have generated thrust only by vertical strokes, which requires an oncoming air flow. This might be achieved either by fast running or jumping from high points or plants while heading into horizontal air currents. While struggling to keep in the airflow, a primitive insect would have swung around into the oncoming current and, flapping its wing-like lobes, gradually descended (see Fig. 4.3, p.68). Vibration of the wing lobes presumably helped the ancestors of winged insects to balance on slender twigs. Modern insects fold their wings during gusts of wind which decreases drag. The short wing pads of primitive insects did not give a streamlined body shape, which is why their owners had to face air currents head on to keep on plants in gusts of wind. Swarming insects still orient themselves this way. It is enough for a primitive insect which sits on the end of a twig and vibrates its wing pads to face the wind, for thrust to be created. All it has to do is to take off and fly forward. An insect flying like this would land quite a long way away from its starting point. Quite soon, because of the flexibility of the wing lobes at their area of articulation with the thorax, simple flapping movements would presumably become accompanied by small rotational oscillations.

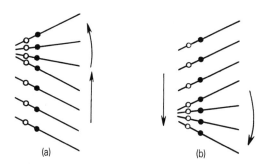

Fig. 5.3 Progressive (straight arrow) and rotational (curved arrow) oscillations in dragonfly wing movement in (a) the upstroke and (b) the downstroke. The centre of mass (filled circle) is behind the torsional axis (open circle) and does not move along the same line. The wing possesses angular momentum around the torsional axis which will cause it to rotate when the wing decelerates.

This only requires the pronational–supinational axis of the wing to shift forwards a little from the centre of mass of the wing, as is the case in the wings of dragonflies (Fig. 5.3). Even such small rotational oscillations made the insects independent of the oncoming wind. By inclining the body relative to the horizontal, they could make way due to the inclination of the resultant aerodynamic force and the production of its vertical component, lift.

This then was probably how the ancestors of winged insects with their short wing-lobes achieved their selective advantage—by increasing their mobility, having more complex behaviour, and becoming capable of travelling great distances. For this to have occurred, even relatively short wing-lobes (less than half the body length) and a small stroke amplitude must have created thrust. Studies of the flight of different insects show that neither the wing length/body mass ratio nor the wing area/body mass ratio limit flight capability. Traumatic shortening of the wings activates compensatory mechanisms and the insect can retain its ability to create lift sufficient to neutralize its body weight (Wilson 1968a). Our data on the flight of mayflies show that *Ephemera vulgata* can even compensate for

its body weight if half of each of its forewings is cut off.

Interesting information about the flight patterns of the first winged insects may be derived from the study of flight of the subimago which, according to palaeoentomological data, was common in many ancient winged insects. The only group in which the subimago stage has been retained up to now is the mayflies. The comparative study of the flight of the imago and subimago of the mayfly *Heptagenia sulphurea* has revealed basic differences in the working of the subimago wing apparatus:

(1) the body angle relative to the horizontal is greater than in the imago (60° as opposed to 0°);
(2) the stroke amplitude is much less than in the imago (87° as opposed to 122°);
(3) the body undergoes strong oscillations in the vertical plane (pitching);
(4) the aerodynamic wake lacks the jetstream *c* (see Fig. 3.14, p.64); i.e. instead of a false vortex street a straight one is produced.

5.2.3 First steps in selection

The first character favoured by natural selection was the lengthening of the primitive wings. Valuable information about flight aerodynamics at this hypothetical stage of the wing apparatus evolution has been obtained by Vinogradov (1959). He studied oscillatory movements in the range of Reynolds numbers most probably used for flight of the first winged insects (10^3–10^4). The maximum Strouhal number (1.0) in the range studied also corresponded exactly to the stroke parameters given above. Under such conditions of the reversed vortex street which follows an oscillating body, three distinct zones are present which vary in degree of stability. The second, most stable zone, which corresponds to the period in which thrust is produced broadens and tends to appear earlier with the increase of oscillations' amplitude. The most effective way of increasing the amplitude of the tip oscillations of the flapping wing is by lengthening the wing itself.

A strong correlation has been found between

the direction of the full aerodynamic force vector which acts upon the insect and the stroke plane angle to the horizontal for a number of different insects: mayflies (Brodsky 1971), the fruit fly *Drosophila* (David 1978), the honey bee (Esch *et al.* 1975), and the small tortoiseshell, *Aglais urticae* (Gewecke and Niehaus 1981). This phenomenon, as well as the characteristic constriction of the wake jet just behind the wings (Brodsky 1988), are consistent with the propeller theory, according to which the most easy way to produce higher thrust, at a given rotational velocity, is to increase the area of the wing disk. This, in turn, increases with increases in wing length and stroke amplitude. Hence any lengthening of the wings would be favoured by selection, and individuals with longer wings would fly more effectively and thus have a better chance of survival relative to other members of the population.

The history of the development of the insect wing apparatus begins with the 'downward fliers', in which natural selection favoured lengthening of the primitive wings and increased flapping movements. As it gave a noticeable advantage in the creation of a force impulse, this step led to a great increase in the inertial forces of the wings. The necessity to overcome these inertial forces determined the next step in selection and the whole course of the evolution of the wing apparatus. The first change was that the adaptive demands of overcoming the increasing inertial forces brought about an asymmetry between the up- and downstrokes.

In recent insects, the kinetic energy of the wing at the end of the up- and downstrokes is absorbed differently. At the end of the upstroke the kinetic energy is eliminated:

(1) by clapping the wings together over the back;
(2) by displacement (deflection) of the thoracic structures (in mayflies the base of the elevated wing deflects the subalar sclerite, in stoneflies, the basalar sclerite, etc.);
(3) by contraction of the muscles used in wing braking (Wilson and Weis-Fogh 1962).

At the end of the upstroke the wing stops for a moment and it is only then that the downstroke

starts. By contrast, the wing moves at maximum speed as it approaches the bottom of the stroke. As the wing stops abruptly, the elastic torsion of the wing base structures is caused by kinetic energy. Inertial forces are transformed into forces of elastic deformation, the latter causing a backward and upward shift in the longitudinal axis of the wing. In consequence, the wing supinates. Thus, the main role in the suppression of the inertial forces at the bottom of the stroke is played not by the thoracic structures and musculature, as at the end of the upstroke, but by the wing itself.

The lengthening of the primitive wings during evolution was accompanied by a growth in inertial forces, the overcoming of which led to the development of supinational deformation at the bottom of the stroke. In their turn, supinated wings are elevated with a smaller angle of attack, and their interaction with the air flow decreases. In Chapter 3 we established that the main function of pronation is to build up the air circulation around the wings at the beginning of the stroke. Following this analogy, we may conclude that the main evolutionary role of supination is to reduce the angle of attack of the wings during their elevation. There is no need to build up the circulation anew at the bottom of the stroke since the stopping vortex plays the role of the starting one. Thus aerodynamic asymmetry between the up- and downstrokes evolved via the following interactions: lengthening of the wings → increased inertial forces → the development of supinational deformation → weakening of the interaction between the wing and the airflow during the upstroke.

The main feature of aerodynamic asymmetry is that the vortex ring produced on the downstroke has a smaller diameter (i.e. it possesses a greater force impulse) than the one generated by the upstroke. The jetstream *c* is formed, which is absent in the mayfly subimago but present in the imago. With a symmetrical stroke, the wings produce a straight vortex street whose axis coincides with the vector of full aerodynamic force, whereas with an asymmetrical stroke a false straight vortex street is generated.

The transition from an aerodynamically symmetrical stroke to an asymmetrical one was an important landmark in the evolution of insect flight. However, we should bear in mind that the process discussed above represents an ideal situation, whereas in reality the first winged insects possessed four wings not two, and consequently their vortex wake pattern was determined more by a phase shift between the wing pairs than by the symmetry of the strokes. We may presume, however, that in the case of synchronous depression and elevation of the two homomorphic pairs of wings, the vortex wake pattern may depend mainly on the way the wings pronate and supinate during the stroke cycle.

As has been already mentioned, mayfly sub-imagos fly at a large body angle relative to the horizontal. The same body position in flight occurs in some caddisflies, primitive moths, and many other primitive insects. This position may reflect the flight conditions of the first winged insects which needed to orientate the straight vortex street at some angle to the horizontal to produce lift. Insects still use a straight vortex street in the first type of hovering, although nowadays this track is produced by a secondary increase in the rotation amplitude of the wing about its longitudinal axis at the extremes of the stroke cycle.

We should also mention the extremely archaic means of producing asymmetrical stroke sometimes used by mayfly subimagos. Edmunds and Traver (1954), who studied the wing mechanics of these insects, suggested that the bullae on the concave veins of their wings caused deflection of the distal part of the wing during the upstroke thus making this half-cycle less active aerodynamically. Our own studies have demonstrated that such strong wing tip deflection does not occur in the imago mayfly. However an abrupt deflection of the distal half of the wing which is retained throughout the upstroke has been *repeatedly* observed in the mayfly subimago.

5.3 Design of the wing apparatus of the first winged insects

The palaeontological history of winged insects starts from the upper Carboniferous (Namurian) (Rasnitsyn 1980*a*). Namurian insects were represented by three clearly distinct groups. The first group includes insects known from fossil records of partial wings and were classified by Sharov (1966) in the order Protoptera. The features of Protoptera which are available for analysis demonstrate their apparent primitiveness, but prevent us from placing them at the base of the phylogenetic tree of winged insects, since their forewings were specialized for a protective function. Insects of the second group belong to the well known order Palaeodictyoptera. Their wings and wing articulations, notwithstanding the antiquity and primitiveness of these insects, were adapted for gliding flight. The third group of insects is represented by the fossil records of wings belonging to two closely related species, *Eugeropteron lunatum* and *Geropteron arcuatum*. Both species presumably belong to a primitive, as yet undescribed order which may belong at the base of the common stem of the Ephemeroptera and Odonata.

Riek and Kukalová-Peck (1984), who described these peculiar wings, consider them to belong to ancient dragonflies from the order Protodonata. Indeed, the pattern of wing venation of recent dragonflies can be quite logically derived from that of the wings of *Eugeropteron* and *Geropteron*. However, the wing venation also resembles that of Permian and contemporary mayflies. The characters of the wings and of the structure of the wing articulation of *Eugeropteron* allow us to put these insects at the base of the common stem of the mayflies and dragonflies, but hardly justify their attribution to only one of these stems. Probably, both species, *Eugeropteron lunatum* and *Geropteron arcuatum*, belong to a very ancient group that was closely related to an unknown common ancestor of the Ephemeroptera and Odonata. This group might be named the Geroptera.

Thus, when describing the organization of the wing apparatus of the first winged insects we should take into consideration details of the wing structure of insects from both the Geroptera and the Protoptera. Protopteran wing structure

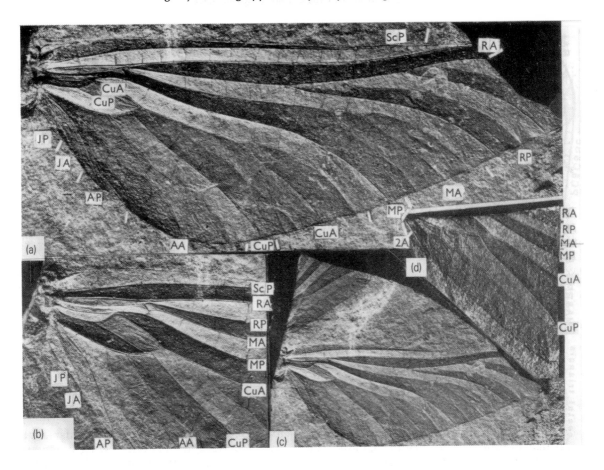

Fig. 5.4 Fragments of the fore- and hindwing of *Eugeropteron lunatum* (after Riek and Kukalová-Peck 1984). (a) Hindwing, length of fragment 35 mm, maximum width 13 mm. (b) The hindwing base under different lighting. (c) The fore- and hindwings in the position in which they were preserved. (d) Forewing, length of fragment 13.5 mm.

is known from the wing fossil records of such species as *Zdenekia grandis*, *Sustaria impar*, and *Evenkia* sp. It must be stressed that many protopterans are known not only from the forewings but also from the hindwings, which is rare for later groups which were capable of locking their forewing in a folded position when at rest. Quite frequently such insects are known only from forewing fragments. *Eugeropteron* and *Geropteron* are known from hindwing fragments, and *Eugeropteron* from a small fragment of forewing too (Fig. 5.4). On the basis of the preserved fragment of forewing we can state that

it had the same structure as the hindwing, but slightly narrower.

5.3.1 Inferences from the wings of the first winged insects

Notwithstanding the apparent differences between the hindwings of the Protoptera (see Fig. 1.19a, p.29) and *Eugeropteron* (Fig. 5.4a), they have some principal features in common. The shape of both wings is not only similar, but rather peculiar. They are triangular, with a drawn out apex and an almost right-angular

tornus. The maximum wing width is one third
of the way from the base. This shape of wing can
easily be derived from nymphal wings in which
the apex is directed obliquely backward, as was
demonstrated by Rasnitsyn (1976). While the
nymphal wings of the Palaeodictyoptera were
arranged in a sequence of presumed ontogenetic
and phylogenetic stages, the wings of the imago
must have developed into the shapes characteristic of the hindwings of Protoptera, the fore- and
hindwings of *Eugeropteron*, and older nymphal
wings of Palaeodictyoptera such as *Tchirkovaea
guttata*. Throughout ontogenesis, the wing pads
must have been transforming with each moult
from a position in which the apex was directed
obliquely caudad to the outstretched position.
That this could occur clearly shows that the insect
could easily move the wings out of the plane in
which they moved during flight. In other words,
they could fold their wings. But how?

In 1966, Sharov suggested that the Protoptera
folded their wings along their back when at rest
with their tips apart as in the fossil imprint of
Sustaria impar. However, it is apparent that the
wings of *Sustaria* were compressed by being bur-
ied, with the right wing displaced and its base
overlapping the left wing. Presumably *Sustaria*,
like some other Protoptera, folded their wings in
a different manner to that suggested by the fossil
record. On the other hand, a simple meeting of
wings of such a shape over the back would
not reduce the width of the body sufficiently.
Compactness of the insect body is achieved by
rotation of the wing in another plane, so that the
leading edge of the folded wing depresses, and
the trailing edge elevates. Thus in the Protoptera,
the folded wings were held in a near roof-like
position, with their tips directed obliquely to
the side.

We can tell how wings are folded by the
structure of the wing articulation. In contemporary insects, wing folding takes place by contraction of the muscles of the third axillary
sclerite; when they contract, the distal arm of
the third axillary sclerite turns over and the
wing folds. Unfortunately, the wing bases of
the Protoptera have not been preserved. In the
wing base of *Eugeropteron* a pair of folds, concave
and convex, are clearly distinct, surrounding a

small (presumably scalped) sclerite designated
here as '3AX' (Fig. 5.5). Topographically, this
sclerite corresponds to the place of attachment
of the movable end, the *insertio* (the immobile
end, the *origo*, is attached to *PWP*), of the
muscle *t–p14* of mayflies and dragonflies (Fig.
5.6). Dragonflies and mayflies have lost the
ability to fold their wings; consequently, the
muscle *t–p14* has changed function. It is strongly
expressed in dragonflies (Newman and Wootton
1988), whereas in mayflies contraction of *t–p14*
brings about remotion of the wing (Brodsky
1974), thus acting in the same manner as in
wing folding. The function of the muscles of
the third axillary sclerite in the Neoptera is well
known (see Chapter 1).

Wing folding in the extant stonefly *Nemoura
cinerea* which was studied by Ivanov (1981), takes
place in several stages. First (Fig. 5.7, 1 and 2)
the wings move upward and slightly backward
while supinated. In the next stage (Fig. 5.7, 3
and 4) the fore- and hindwings are shifted to
the mid-line of the body, and the hindwings

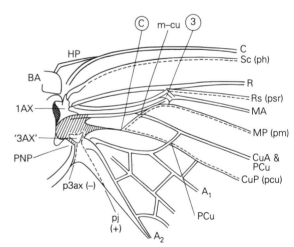

Fig. 5.5 Hindwing base of *Eugeropteron lunatum*,
an interpretation from the photo shown in Fig.
5.4. The concave veins and folds *Sc, Rs, MP*, and
CuP are indicated by continuous and broken lines
running in parallel. 3 is the torsional hinge, point
C corresponds to the apex of the distal arm of the
third axillary sclerite.

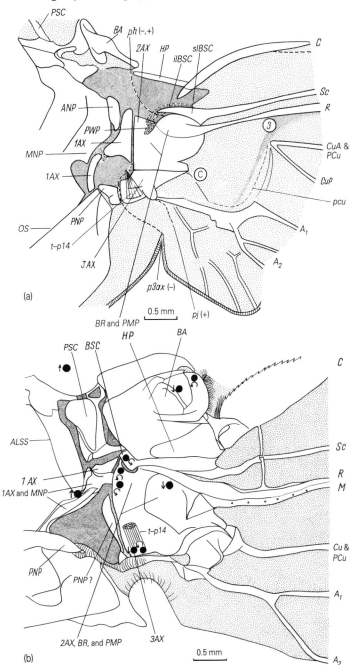

Fig. 5.6 The forewing bases of (a) the mayfly *Siphlonurus immanis* and (b) the dragonfly *Aeschna juncea*. In both cases, the mobile tip of the muscle *t–p14* that flexes the wing is attached to the plate at the posterior part of the wing base which comprises fused *3AX, 2AX, BR*, and *PMP*. Solid spots in (b) correspond to the sites of attachment of the flight muscles, with an indication of their action by straight (wing depression and elevation) or curved (pronation and supination) arrows.

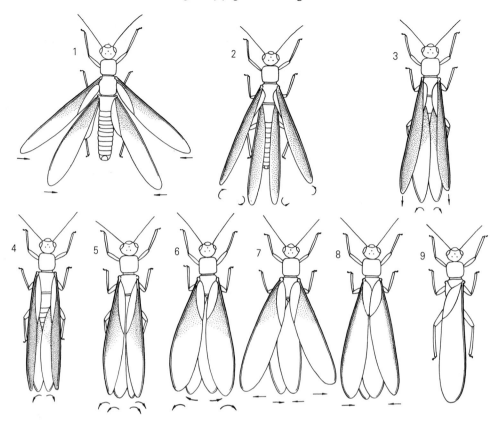

Fig. 5.7 Consecutive stages of wing folding in the stonefly *Nemoura cinerea*, as seen from above (after Ivanov 1981). The direction of movement of the wing tips is indicated by arrows. The complete wing folding process includes several stages, as follows: (1) abduction; (2–4) supination; (5–8) pronation; and the final settling on the abdomen (9). The wings of the first winged insects are supposed to have taken up a position close to that shown in (6). The wings take this position when the third axillary sclerite is incompletely turned over or when its distal arm is short.

begin to pronate. Then the forewings pronate (Fig. 5.7, 5 and 6) and the hindwings, turned horizontally and lying flat over the abdomen, become completely pronated and move slightly in the frontal plane. In the last stage (Fig. 5.7, 7 and 8) the hindwings are folded, and the pronated forewings, moving in the frontal plane, are brought close to the side of the body and finally folded over the abdomen (Fig. 5.7, 9). The hindwings clearly pre-empt the forewings. The position of the folded wings in the first winged insects can be presumed to have been close to the one in which the fore- and hindwings are fully

pronated but are not yet quite on the middle line of the body (Fig. 5.7, 6).

The position of the resting wings of the Carboniferous-Permian Diaphanopterodea and Neoptera (Fig. 5.8) was similar to that described above. In both, unrelated, groups the wings are primitively folded in an oblique-lateral position. Later on the wings gradually became flatter and crossed over in many lineages of the Neoptera. The adaptative advantages of the changes in wing folding were very probably the stream-lined body which allowed the insect to hide in narrow spaces, more secure locking of the wings

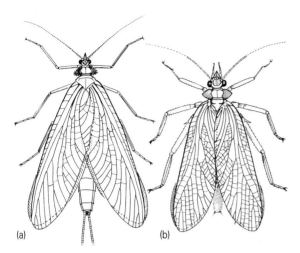

Fig. 5.8 Reconstruction of living animals, showing the positions of the folded wings. Missing structures are hatched (after Kukalová-Peck and Brauckmann 1990). (a) *Namurodiapha sippelorum*, an early Late Carboniferous insect of the order Diaphanopterodea. The wings probably overlapped in the anal area. (b) *Heterologopsis ruhrensis*, an early Late Carboniferous insect from the hemipteroid assemblage. The wings are folded in the oblique-lateral position without crossing over.

in the folded position, and better protection of the folded wings against mechanical damage.

It may therefore be concluded that wing folding in the first winged insects was powered by contraction of the muscle *t–p14* by turning '3AX' over, but that its distal arm articulating with *PCu* and *A*, was not long enough to move the wings to the mid-line of the body.

In addition, taking into account the peculiar shape of the hindwings of the first known winged insects, we can draw inferences about their deformational patterns in the course of a stroke cycle. As has been shown by Brodsky (1987), wing deformation during supination influences its structure to the greatest extent. Supinational deformation of the wing is in turn determined by the wing shape. For wings of the shape discussed above, in which the veins are alternately concave and convex, gradual supination is habitual. This pattern of wing deformation

is typical of most mayflies. Before we start to discuss the process of gradual supination, it should be noted that an alternation of concave and convex veins is also habitual in the hindwings of the Protoptera, although to a lesser extent. For example, in *Zdenekia* the following alternation of veins is observed: $C+$, $Sc-$, $R+$, $Rs-$, $M+$ (before $m-cu$), MA? (here the wing is suppressed, so it is difficult to identify the original position of the vein on the membrane), $MP-$, $CuA+$, $CuP-$, and $PCu+$, and the other veins occupy a neutral position on the wing membrane. Deformation of the mayfly *Ephemera vulgata* wing during the upstroke occurs as follows (Brodsky 1981): 'The wing stops abruptly and starts to supinate: first the leading edge of the wing lifts causing oblique bending of the wing along the concave vein Rs_5 (Fig. 5.9a), then the bending shifts to another concave vein MP_1, then to MP_2, until finally only one line of bending, the concave vein CuP remains. This stays until the end of the supination phase, at which point the wing plane becomes vertical, with its leading edge oriented upward'. It should be added that in mayflies longitudinal bending of the wing surface starts with a slight wing tip deflection first downwards, then upwards. The other details about wing supination in *Eugeropteron* are presumably exactly the same as in mayflies (Fig. 5.9b). Their torsional hinges, and probably the torsional hinge of the protopteran wing which is not preserved in fossil wings, are presumably homologous as well.

The specialization of gradual supination which is, as has been already mentioned, a consequence of wing shape, has led to a number of further adaptations. First, a strict alternation of concave and convex veins is necessary to allow gradual supination, and second, a torsional hinge is formed. During supination, longitudinal bending of the wing occurs along concave vein-folds that converge at the wing base between M and PCu, level with the apex of the future distal arm of the third axillary sclerite (see Fig. 5.5, p.90). The development of the lines of wing bending gave rise to considerable modification of the venation pattern which constitutes framework of the wing. At the point of intersection of the medial fold and the transverse vein $m-cu$, a curious curve in CuA appeared, and the transverse vein itself became an accessory torsional hinge. The situation presumably led to both of these features of the wing structure—the presence of the transverse vein *m–cu* and the curve in *CuA*—becoming stable and typical of most insects, including the Palaeodictyoptera.

One more conclusion can be drawn from the

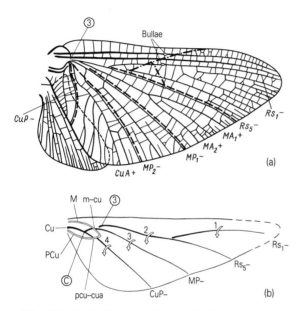

Fig. 5.9 Specializations for gradual supination in the wings of two different insects. (a) The fore- and hindwings of the mayfly *Ephemera vulgata*. Lines of supination are indicated by dashed lines, and the line of wing tip deflection with a dot-dashed line. (b) Schematic diagram showing how gradual supination during the upstroke was accomplished in the wing of *Eugeropteron lunatum*. The whole wing surface is divided into sectors by concave vein-folds. During supination, each sector turns around a posteriorly located vein-fold in the direction indicated by arrows. Supination starts from the leading edge and proceeds towards the rear edge in the order indicated by the figures. Each of the vein-folds forms a kind of longitudinally oriented hinge. The hinges of the fore part of the wing are concentrated in the torsional hinge 3 (circled). The bases of the vein stems *M* and *PCu* constitute the main supporting structure, which is crossed by the concave vein-fold *Cu*.

fossil records of the wings of the first known insects. In a number of her papers, Kukalová-Peck (1983, 1987, and others) developed the hypothesis that the structure of the wing articulation of the first winged insects was a zone consisting of 32 plates aligned in eight rows which provided a robust join between the wing and the tergum. This hypothesis is based on the study of the well preserved wing articulation of the palaeodictyopteran *Mazonopterum wolfforum*, as well as on the uniformity of wing articulations in other species of Palaeodictyoptera. The wing articulation areas of palaeodictyopterids consisted of heavy 'platforms' from which the vein stems started in a parallel row (Figs 5.10b and 5.11). A wing with this sort of base construction is well adapted for gliding, but its ability to rotate about its longitudinal axis during supination is limited. In all cases in which the articulation zones of the wings are well preserved, it is clear that the bases of numerous anal veins are cut by a deep fold whose upper end comes close to the torsional hinge (Figs 5.10 and 5.11). The torsional hinge consists of a bundle of curved veins with a large membranous field beneath (Fig. 5.10a). This is obviously the only way to provide twisting in a system consisting of parallel supporting structures. The fold mentioned above crosses the anal veins, *PCu* and enters the cubital field, but never crosses the cubitus, which makes it like the jugal fold (see Figs 5.5 and 5.11). *Rs*, *M*, and *Cu* are curved and converge in the area of the torsional hinge; furthermore, the shape of the curve of *M* and *Cu* is exactly the same as the structure of the torsional hinge in *Eugeropteron*: the curved part of *Cu* lies more proximally than the corresponding part of *M*, and both veins are concave on the wing membrane in the hinge area (Figs 5.5 and 5.10a). All this enables us to derive the wing base structure of the Palaeodictyoptera from a relative of *Eugeropteron*, but not vice versa. It should also be emphasized that in all cases the development of gliding flight was accompanied by the development of platforms in the wing bases which provided a robust link between the wing and the thorax; this process occurred independently in at least three groups: the Odonata, diurnal Lepidoptera, and the Palaeodictyoptera (Brodsky 1989).

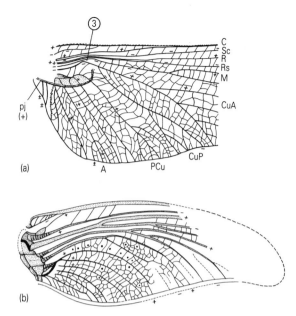

(a)

(b)

Fig. 5.10 Fragments of palaeodictyopteran wings showing (a) the structure of the torsional hinge and (b) a platform at the wing base. (a) base of hindwing of *Thesoneura americana*, total length 110 mm. Note the line of weakness that crosses the bases of the anal veins and *PCu* on its way to the torsional hinge (after Riek and Kukalová-Peck 1984). (B) Wing of a young primitive nymph of *Parathesoneura carpenteri*. Note that the basal sections of wing veins other than *R* were flexible because they were formed of alternating strips of thicker and thinner cuticle, perhaps to provide additional mobility, which was reduced by the fusion of the vein bases into a common integrated platform.

Thus, poor development of the torsional hinge in palaeodictyopteran wings clearly demonstrates that the extent of rotation of the wing about its longitudinal axis is too low to provide normal large amplitude strokes. The pattern of wing articulation of the first winged insects should be sought not in the Palaeodictyoptera but in insects which flap their wings with a relatively large amplitude. This in turn implies the presence of soft wing articulation. In *Eugeropteron* (Fig. 5.5), the first axillary sclerite is located in the

wing base between the anterior vein stems and the anterior notal wing process. The posterior veins come close to the well developed posterior notal wing process, and the rest of the wing articulation is soft. An imaginary line connecting the first axillary sclerite, the torsional hinge, and the posterior notal wing process delimits a roughly triangular soft deformable zone; its external border is formed by the concave vein-fold *MP* (*pm*) (Fig. 5.12). There are no large sclerites or platforms in the deformable zone.

5.3.2 Structure and function of primitive wing apparatus: a summary

The first winged insects presumably had two homomorphic pairs of wings with characteristic triangular shape and a drawn-out apex. The insects flapped their wings in synchrony, with a large amplitude, as is demonstrated by their well-developed torsional hinges. The clearly distinct folds *psr*, *pm*, and *pcu*, which ran along the longitudinal veins and rendered them concave on the membrane, permitted gradual supination

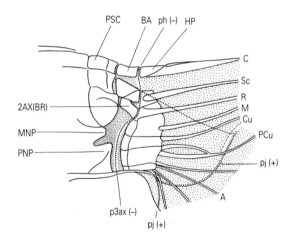

Fig. 5.11 Interpretation of the wing base of the palaeodictyopterid *Mazonopterum wolfforum*, based on drawings in the work of Kukalová-Peck and Richardson (1983). The fold that crosses the anal and postcubital vein bases seems to be a part of imaginary fold system (dotted lines) which allowed wing folding.

of the wings during the upstroke. The wing articulation was soft, with a distinct first axillary sclerite at which the bases of the anterior veins converged. Wingstrokes which have a large vertical amplitude relative to the longitudinal body axis require soft wing articulation with few but effective hinges. The horizontal hinge located between the first axillary sclerite and notal margin forms one of these. There was no vertical hinge controlling wing movements in the plane perpendicular to the wingstroke. The wing venation was partially specialized; for instance, the presence of the torsional hinge influenced the anastomosing and curving of the bases of the medial and cubital veins.

When at rest, the wings were held backwards and arranged in a roof-like position: the hindwings were directed with their tips backwards and sidewards with a depressed leading edge, and the forewings covered them to some extent, the tips also pointing backwards and sidewards. The forewings were not locked in the folded position, so that the wings of dead insects were often displaced. This probably explains why protopteran hindwings are relatively frequent in the fossil record, whereas fragments of forewings only occur commonly for insects with locked wings. The folding of the wings was produced by the muscle *t-p14*, by overturning a small sclerite located between the posterior notal wing process and the bases of the posterior veins. Both segments of the pterothorax were developed equally as were the basalar and subalar sclerites to which

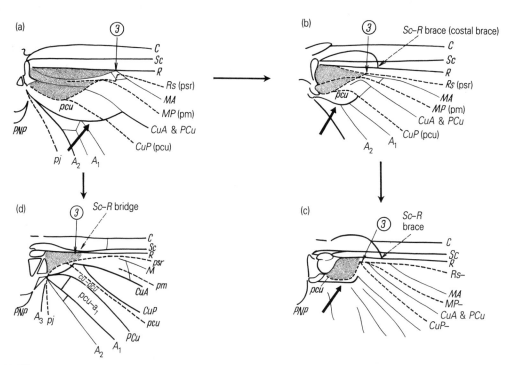

Fig. 5.12 The structure of the wing base in two evolutionary lines. The soft deformable zones, roughly triangular in shape, are stippled. Solid arrows indicate a postcubital brace in the ephemeropteran evolutionary lineage. The reasons for evolving a costal brace are discussed elsewhere (Brodsky 1970*b* and 1987); the costal brace serves to shorten the costal edge of the wing during the downstroke rather than to strengthen it. (a) *Eugeropteron*-type; (b) Permoplectoptera-type; (c) Recent Ephemeroptera-type; (d) Neoptera-type.

the direct flight muscles were attached. As a consequence of the symmetrical development of the direct flight musculature the pleural suture ran vertically rather than obliquely. The terga of the pterothoracic segments lacked scutellum and consisted of simple plates with a few sutures. Furthermore, a comparison of the musculature of the wing-bearing segment and the abdominal segment of mayfly nymphs which have tracheal gills indicated that the musculature of the wing-bearing segment of the first winged insects was complete. It included, in addition to the basalar and subalar muscles, the following groups of muscles: tergal, tergosternal, tergocoxal, tergo-trochanteral, and tergopleural muscles.

5.4 Three main evolutionary lines in the history of winged insects

The axillary apparatus of modern insects is a complicated system of sclerites, bases of veins, and folds, in which the sclerites move relative each other by means of joints and hinges. Sclerotization of the wing bases must have occurred at some stages in the evolution of the wing apparatus. The process of sclerotization was provoked by the necessity to fix the wings either outstretched (for gliding) or when folded, and also to allow complex wing movements during strokes. In other words, soft wing articulation, which was only able to allow strokes in the vertical plane and imperfect folding, sooner or later must have given way to a system of joints and hinges. Functionally important folds and articulations which already existed determined the shape and location of the sclerites of the axillary apparatus. The leading part was played here by the vein-folds, which provided deformation of the wing plane during supination: *psr*, *pm*, and *pcu*. These three folds meet at the point C in the *Eugeropteron* wing (Fig. 5.5) which corresponds to the apex of the distal arm of the third axillary sclerite of the generalized axillary apparatus and from which a deep split broadening toward the notal margin originates. Besides these folds, the jugal fold and the fold of the third axillary sclerite already existed, providing primitive folding of the wing.

If we add sclerites of the wing base to the existing system of folds (Fig. 5.13), then to reach the generalized axillary apparatus it would be quite enough to assume that the detached bases of the postcubital and anal veins were taken into the third axillary sclerite. With the acquisition of the distal arm, the third axillary sclerite became much more effective, so that contraction of the muscle *t–p14* now brought the folded wings closer to the middle line of the body. The jugal fold and the fold of the third axillary sclerite met at the base of the radius (Fig. 5.13), causing a vertical hinge to be developed. The wing acquired one more degree of freedom, its longitudinal axis became able to move forward on the downstroke and backward on the upstroke.

The process of sclerotization of the wing articulation occurred concurrently with the development and accumulation of apomorphic changes

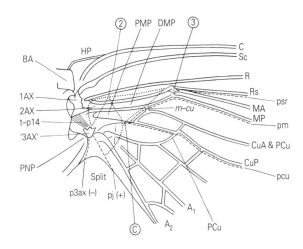

Fig. 5.13 Generalized scheme of the axillary apparatus, obtained by 'superimposing' sclerites onto a system of folds which already existed at the wing base of *Eugeropteron* (Fig. 5.5). *i* and *j* are detached bases of *PCu* and *A* which constitute the lobes of distal arm of the third axillary sclerite. The borders of sclerites appearing later are shown by a dotted line.

in each phylogenetic lineage. By the end of the process of the wing base sclerotization, the phylogenetic lines had diverged significantly (Fig. 5.14). We do not know how and when the three main lines of evolution of winged insects diverged but it is obvious that all of them inherited the basic principles of the organization of the wing apparatus. Beginning with the common wing base pattern, the process of wing base sclerotization proceeded separately in the different lineages.

It has been emphasized repeatedly (Brodsky 1988, 1989) that the axillary apparatuses of mayflies and dragonflies are based on the same pattern, which differs only in details. Moreover, Rasnitsyn (1981) pointed out the lack of significant differences in the wing articulation structure between the Diaphanopterodea and the Neoptera. When considering the changes in function and structure of the wing apparatus in each of the lineages at the divergence stage and at the stage of the sclerotization of the wing articulation, it should be remembered that all existing patterns of axillary apparatus, both recent and extinct, can be derived, as is demonstrated in Fig. 5.13, from a wing base close to that of *Eugeropteron* but lacking anastomosis between *PCu* and *CuA*. Different types of wing apparatus have evolved in the course of evolution. We agree with Kukalová-Peck and Brauckmann (1990) that '. . . the transformation of wing venation was a mosaic, and the rate of this transformation differed among the groups, long before the Pterygota were first recorded in the early Late Carboniferous'.

5.4.1 The divergence stage

Mastery of the airspace by the first winged insects was a breakthrough into a completely new adaptive zone. What vacant niches were discovered there by the descendants of the first winged insects? To examine those niches one needs to imagine the close, moist atmosphere of Carboniferous swamps. Tropical woods, with luxuriant flora of tree lycopods, psilopsids, ferns, pteridosperms, and cordaitopsids which covered the banks of lagoons and shallow seas, sheltered myriads of plant-feeding insects. Strange insects

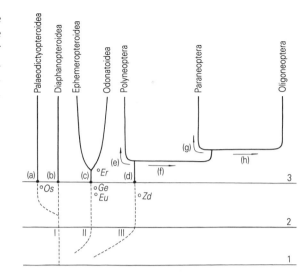

Fig. 5.14 Basic phylogenetic scheme of winged insects showing three stages of apomorphic changes: 1, common ancestor; 2, divergence of the three evolutionary lines (Palaeodictyopteroidea and Diaphanopteroidea (I), Ephemeropteroidea and Odonatoidea (II), Neoptera (III)); 3, existence of palaeontological evidence of wing base sclerotization for the four phylogenetic lineages (a, b, c, and d). The apomorphic characters for each lineage include: (a) gliding, by fusion of the *Sc-A* vein bases; (b) wing twisting, by fusion of the *R* and *M* vein bases; (c) operation of the wing by *PNP*, with the consequent development of the *SA* muscles, via the *PCu-CuA* bracelet; (d) wing operation by *ANP-MNP*, with the consequent development of the *BA* muscles, wing folding by *3AX* by means of the long distal arm, using A_3, and elytrization of forewings. The arrows e–h show the direction of apomorphic changes in Poly-, Para-, and Oligoneoptera: (e) wing operation by *ANP*, the development of *p-SCT*, and hindwing expansion by A_2 branching; (f) development of a true *SCT* with *t13* and wing folding using both *t–p14* and *t–p13*; (g) wing operation by fused *ANP* and *MNP* and wing folding by reduction of joint B; (h) wing operation by *MNP* and wing folding by hinge-lever-stopping systems and development of *LPN* etc. *Eu, Eugeropteron lunatum; Er, Erasipteron larischi; Ge, Geropteron arcuatum; Os, Ostrava nigra; Zd, Zdenekia grandis;* relative dating according to the fossil records.

with big spotted wings and a very small head armed with a long hard proboscis formed almost half of the Upper Palaeozoic entomofauna. With their mouthparts modified for piercing and sucking they tore off the loosely constructed cones of the tree lycopods and Cordaites and drilled seeds and megaspores, sucking out their contents. Palaeodictyopterids and diaphanopterids were abundant in the crowns of tall plants, where the adult insects and their nymphs spent all their time. The protopterans made up another large group of plant-feeding insects, which, with their chewing mouthparts, fed on the generative organs of plants. They could partly use the falling pollen as well, gathering it from the surface of the substrate as do present-day neuropterans and scorpionflies. While doing this they spent a great deal of time on the ground among the detritus of vegetation, in which they could hide quickly when predators appeared. The insects only avoided open spots at forest edges and above water spaces in which, as inhabitants of tree crowns, they could not withstand gusts of wind, and where protopterans at ground level were unable to find safe shelter. It was apparently in these sites that the ancestors of present-day mayflies and dragonflies started to become established.

Habitats such as the crowns of trees which were inhabited by insects of the first evolutionary stem (Fig. 5.14, I) could be exploited in two different ways, depending on the insects' body size. Compared with the robust Palaeodictyoptera which had wingspans of up to 560 mm (Kukalová-Peck and Richardson 1983), the Carboniferous Diaphanopterodea were relatively small, with a wingspan of 70 mm. These smaller insects were presumably adapted to fly in the limited spaces in the dense vegetation which were inaccessible to the large palaeodictyopterids. It is difficult for large insects to manoeuvre among dense vegetation, and they cannot shelter in crevices. The large insects must have flown among, and perched on, the pole-like trunks of the coal swamp trees, feeding on the long, pendulous cones of lycopods and cordaitopsids and on the exposed ovules of seedferns (Fig. 5.15).

The ancestors of the second evolutionary stem

(Fig. 5.14, II), that gave rise to the Ephemeroptera and the Odonata, had to adapt to living in the exposed spaces in which, first, the likelihood of being blown about was considerable and, second, the areas containing food resources and mating and oviposition sites were dispersed. This probably resulted in the evolution of the different modes of territorial behaviour known as swarming flight in mayflies and patrolling in dragonflies. Gusts of wind compel insects to intensify their flight so as to maintain a constant position in space. For example, when a swarming mayfly male tries to keep over a marker in a strong wind, the character of its wing movements changes: the stroke

Fig. 5.15 An early Permian palaeodictyopterid *Goldenbergia* sp., reconstructed feeding on a megaspore (after an original drawing by Ponomarenko 1980).

amplitude increases, the path drawn by the wing tip shifts backward, and the speed of wing lowering increases (Brodsky 1975). Such changes in wing kinematics are caused by the subalar muscles taking a more active role in wing depression. Presumably, at one stage in the development of the wing apparatus of the ancestors of mayflies and dragonflies, the musculature of the subalar sclerite had the main role in wing depression so that control of the wing occurred via the posterior notal wing process which was well developed even in *Eugeropteron*. The principal aerodynamic pressure during the downstroke is on the leading edge of the wing, whereas the work of the wing depressors is exerted on the wing via the posterior notal wing process and consequently on the trailing edge of the wing. This is a contradiction that can be resolved as following: the postcubitus turns and joins *CuA* (Fig. 5.12a).

The development of the anal brace (or more precisely the postcubital brace) may be assumed to be the basic synapomorphic character of the Ephemeropteroidea and the Odonatoidea; its evolutionary path, as well as that of the roughly triangular soft deformable zone can be traced through the fossil records from *Eugeropteron* right up to present-day mayflies (Fig. 5.12a–c). By means of the anal brace, the work of the wing muscles is transferred to the wing leading edge distally from the torsional hinge. A structure analogous to the postcubital brace of mayflies is also present in the Neoptera (Fig. 5.12d), but in this group it developed from remnants of archedictyon into the transverse vein *cu–pcu* (see Fig. 1.13, p.22). This *cu–pcu* varies in shape between individuals of the dobsonfly *Corydalis* sp.: it is often Y-shaped, but sometimes the left or right branch may be absent.

The third phylogenetic stem comprises most winged insects, including members of the Protoptera (Fig. 5.14, III). The forewings of the Protoptera apparently evolved in the direction of elytrization; a lanceolate shape, smooth corrugations, and the rich network of the archedictyon were general in the group. Elytrization, in turn, is only feasible if the folded wings are stable. The evolutionary success of this group was due to an optimal combination of advantages in flight

ability and improvement in wing folding. It is likely that the protopteran wing base already had A_3 functioning as a strut of the deflecting lobe during wing folding. The improvement in flight led to the development and strengthening of the muscles of the basalar complex. This means of improving flight was more productive than that adopted by the insects of the second evolutionary branch, as the main aerodynamic pressure, as has already been mentioned, is on the leading edge of the wing. Consequently, the main role in wing functioning was taken by the anterior region of the notal margin and the first axillary sclerite. This optimal combination of improvements in both flight and wing folding enabled the insects of this evolutionary branch to hide in confined spaces, to use flight to find such places, and to find dispersely located feeding and oviposition sites as well.

5.4.2 The wing base sclerotization stage

Sclerotization of the wing bases occurred concurrently with the modification of the wing apparatus. Once it was complete, all the previous modifications seem to have been 'recorded' at particular locations in the sclerites of the axillary zone.

The Diaphanopterodea, of the first evolutionary stem, retained the ability to fold their wings allowing them to make use in a different way of the environment dominated by the large, slow, and short-distance flying Palaeodictyoptera. In one of the primitive diaphanopterids *Martynovia protohymenoides* the wing base was traversed by well pronounced folds that probably played a part in wing folding (Fig. 5.16). A deep jugal fold separated the extensive field of sclerotization from the wing surface. This fold arose from the posterior margin of the wing, projected forward and, as in the Palaeodictyoptera (Fig. 5.11) and the probable ancestors of the Ephemeroptera and Odonata (Fig. 5.13), cut across the bases of all the veins in its way. The humeral fold, which was continuous with the jugal, abruptly turned to the side of the wing articulation just in front of the subcostal vein, crossed *Sc* and ended by the anterior margin of the first axillary sclerite. The sclerotization of the wing base was crossed

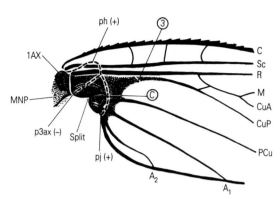

Fig. 5.16 Forewing base of the diaphanopterid *Martynovia protohymenoides* (after Kukalová-Peck 1974, with altered labels). A characteristic split is shown in the wing base, from the end of which (point C) the base of the cubital vein begins in the *Eugeropteron* wing base (Fig. 5.5).

by another fold, that of the third axillary sclerite. Kukalová-Peck (1974) considers that this fold is convex. If this is the case, the mechanism of wing folding in the diaphanopterids was completely different from that of the Neoptera, because in the latter this fold is concave, whereas the jugal is convex. However, it is quite difficult to recognize the nature of folds from fossils, and this needs to be taken into account when considering the functions of the axillary apparatus during wing folding and extension. In more advanced diaphanopterids, for instance in *Permodiapha carpenteri*, the subcostal vein curves at its base, providing room for the humeral fold. This therefore supports the belief that wing folding in the Diaphanopterodea was basically the same as in the Neoptera and, notwithstanding the new data on the wing base structure which has appeared recently (Kukalová-Peck and Brauckmann 1990), attention should be paid to the close resemblance between the wing base of *Martynovia* (Fig. 5.16) and *Eugeropteron* (Fig. 5.13).

A decrease in the distance between the $R+M$ and Cu veins occurred in the diaphanopterid lineage (Fig. 5.14b), and a characteristic medial plate was formed outside the jugal fold, on whose apex there was a torsional hinge (Fig. 5.16). The wing stroke amplitude was rather small in

view of the structure of this hinge; less than in *Eugeropteron*, but considerably larger than in the palaeodictyopterids. Their smaller body size, their ability to fold their wings, and their swifter flight which was provided by the two pairs of almost homomorphic wings allowed the diaphanopterids to withstand the competition from the Palaeodictyoptera.

In the first evolutionary stem a branch of gliding forms (Fig. 5.14a) diverged from the Diaphanopterodea, due to their large body size and feeding habits. These palaeodictyopterids fed on the endosperms of seeds, sucked out by their piercing beaks (Sharov 1966). It is likely that this mode of feeding required extra-intestinal digestion which, in turn, compelled the insect to spend a long time immobile at one site. While stationary and exposed, a large insect is at a high risk of predation. Thus the palaeodictyopterids had to be ready to drop down at any moment to elude their enemies. Since endosperms were abundant, there was no need for lengthy migratory routes to find unexploited food resources. Short flights among plants searching for endosperms, egg-laying sites, or escaping from enemies could be achieved as well by gliding, an energy-saving and ever-ready flight mode. Their large size and long cerci which helped to support the tip of the abdomen during flight also favoured gliding. Having large gliding wings they required more branching veins to support the wing membrane but were not subject to the autapomorphic adaptations linked with more laborious flight and with wing folding. Once gliding became overwhelmingly important the ability to fold the wings degenerated, and with time the palaeodictyopterids lost the ability to fold the wings entirely.

The wings of the adult palaeodictyopterids were permanently horizontally outstretched because of secondary fusion between several articular sclerites and the wing veins. At the wing base the areas among the vein bases became sclerotized, producing a platform that provided a strong link between the outstretched wings and the thorax. The platform was formed by the fusion of all the vein bases from the subcosta up to the anal veins (Fig. 5.11). The ability to fold the wings was lost, the jugal fold changed direction and began to participate in the work of the torsional hinge, since the strokes were retained even though the amplitude decreased.

Gliding played an important part in the flight of insects of the second evolutionary stem (Fig. 5.14c) too. When an insect has to spend a long time in flight the problem of saving energy becomes real. Gliding flight was usual in the Meganeurida and still occurs in the anisopterous Odonata and recent mayflies. It is not known whether this adaptation evolved in the common ancestor of the Odonata and Ephemeroptera or appeared later, after these lineages had diverged. Unlike dragonflies, which hold their wings in a gliding position without any muscle effort (Neville 1960), in mayflies the wings are held outstretched by the tonic contraction of a number of tergosternal muscles (Brodsky 1974). Anyway, wing base sclerotization in both dragonflies and mayflies was greatly affected by the wing operating via the posterior notal wing process. Consequently, the posterior area of the wing articulation became the centre of sclerotization. In mayflies and dragonflies the platform which strengthens the wing base and connects it with the tergum developed as a result of fusion of the basiradiale, the second axillary sclerite, the proximal medial plate, and the third axillary sclerite (Fig. 5.6)—it is not homologous with the platform in the wing base of the Palaeodictyoptera. Hence the development of the wing apparatus of insects in the second evolutionary lineage was guided by adaptations for gliding and consequently by the loss of the ability to fold the wings. This occurred from the starting point of wing operation via the posterior notal wing process and the accessory development of the muscles of the subalar complex.

Flight and wing folding improvements need to be considered separately in the third evolutionary lineage (Fig. 5.14d). The wings became lighter and the rich network of archedictyon was gradually reduced leaving numerous transverse veins. The folds of deformation shifted backward from the longitudinal veins (Fig. 5.12d) so that the corrugations in the wing were smoothed out, and wing deformability increased. The distance between the torsional hinge and the wing base became shorter. In the region where the folds crossed vein bases, the roots of M and Cu became less substantial, thus increasing the wings' capacity for supinational deformation. The wings were folded more securely, enabling the insect to use confined spaces for hiding. The vertical hinge became fully developed, and the transverse veins cu–pcu and pcu–a_1 formed away from the jugal fold as a result of a reduction of the archedictyon. These short veins connect the ends of vein stems, providing the surface of the folded wing with rigidity. The elytrization of the forewings was accompanied by changes in their deformation: gradual supination was replaced by supinational twisting (which will be discussed later). The hindwings still underwent gradual supination during the upstroke.

6

Early forms of flight

Today I saw the dragon-fly
Come from the wells where he did lie.
An inner impulse rent the veil
Of his old husk: from head to tail
Came out clear plates of sapphire mail.
He dried his wings: like gauze they grew;
Thro' crofts and pastures wet with dew
A living flash of light he flew.

Alfred, Lord Tennyson

The first stage in the evolution of the wing apparatus, the transition from aerodynamically symmetrical to asymmetrical flaps, ended with the appearance of actively flying insects. The stages of evolution that followed were concerned with the optimization of the interaction between the wing pairs and with an increase in wingbeat frequency. This trend is observed in the evolution of several different lines of winged insects.

6.1 The departure from synchronous working by the wing pairs

Selection of the optimal phase relationships for wing pairs movement apparently started from a synchronous pattern, in which both pairs of wings moved with an equal and large amplitude. In this situation, the hindwings apparently pass through the vortex twists flowing off the forewing tips, which causes changes in the velocity fields of the hindwings (Fig. 6.1). A diminution in the negative influence of the vortex twists crossing the hindwings could be achieved by either a break in the synchronicity of the wing pairs, or by a decrease in wing stroke amplitude; alternatively, the hindwings may be shortened. Insects that evolved longer wings and a decreased stroke angle were able to retain the initial synchronicity of the wing pairs. On the basis of the structure of the wing and of its articulation, it is possible that flapping flight in Palaeodictyoptera consisted of extremely low amplitude strokes (see Fig. 4.3, p.68). The forewings broadly overlapped the hindwings producing strictly synchronous flaps. These insects presumably learned to use their large size at an early stage to transform a fall-jump from a plant into gliding flight. Their wings, as well as those of the Meganeurida, are long, with extended tips, and are perfectly adapted for gliding (Fig. 6.2).

It was not only the insects with a small stroke amplitude (Palaeodictyoptera, anisopterous Odonata, and Meganeurida) which had equal length wings. It is also the case in insects that move their wings with a large phase shift between the pairs: the zygopterous Odonata and some Megasecoptera (Fig. 6.3a). Both these groups are characterized by a petiolate wing form and narrow wing bases. These insects presumably acquired an increase in stroke amplitude secondarily,

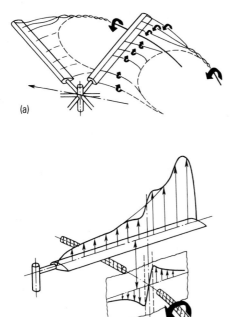

(a)

(b)

Fig. 6.1 The influence of vortex twist from the forewing tip on airflow around the hindwing. (a) General view of the vortex twist flowing off the forewing tip, and (b) the velocity field of the hindwing. The solid arrow indicates the direction of rotation of the vortex twist; thin arrows indicate the velocity field of the hindwing (adapted from Wagner 1980).

but since the stroke planes of the fore- and hindwings are inclined differently relative to the horizontal, the negative influence of the vortex twists flowing off the forewing tips did not affect the airflow around the hindwings. In cases where insects retained their original wingstrokes synchronicity, with an increased amplitude, the effect imposed by the vortex twists on the hindwings grew and led to their noticeable shortening (as in some Megasecoptera (Fig. 6.3b)), right up to complete reduction in the Permothemistida and Eukulojidae. Thus, by the Permian, there were two different groups of Palaeodictyopteroidea, the Permothemistida and the Eukulojidae, in which dipterigy (two-wingedness) had arisen.

Reduction of the hindwings caused diminution of the metathorax, and concentration of the flight muscles in the mesothorax caused it to develop intensively. In addition, the wing became costalized; expressed by the reinforcement of its leading edge and a reduction in the venation in the rear part of the wing plate. Outwardly, the Permothemistida were quite similar to mayflies and in 1949 Rohdendorf had already suggested that the flight of these insects may have resembled that of modern mayflies. The Permothemistida were markedly different from mayflies in the extent of development of their antennae: these are short in mayflies, but 2.5

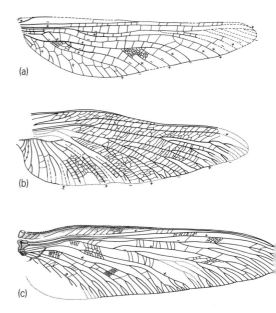

(a)

(b)

(c)

Fig. 6.2 Wings of some very early insects. (a) The meganeurid dragonfly *Erasipteron larischi*, forewing fragment: length 55 mm, total length c. 64 mm (after Riek and Kukalová-Peck 1984). (b) The palaeodictyopterid *Larryia osterbergi*, hindwing of young flying subimago: length 87 mm (after Kukalová-Peck and Richardson 1983). (c) The palaeodictyopterid *Ostrava nigra*, forewing: length 74 mm. Note the remains of the archedictyon between the longitudinal veins and also the platform at the wing base (after Kukalová-Peck and Richardson 1983). These elongate wings with sharpened tips and a broadened base were beautifully adapted for gliding.

times the body length in the Permothemistida. The lengthened forelegs of mayflies, which are stretched out forwards when parachuting during the mating dance, presumably serve as an alternative to having long antennae.

The structure of the wing apparatus of the Eukulojidae differed greatly from that of the Permothemistida in many apomorphic features:

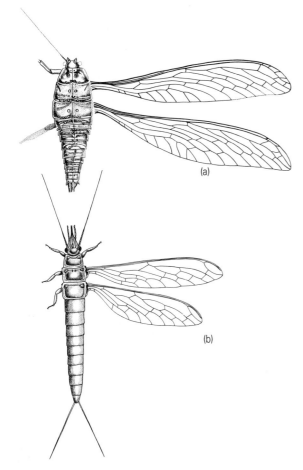

Fig. 6.3 Reconstruction of megasecopterid insects, with the right-hand-side fore- and hindwings extended. (a) *Sylvohymen sibiricus*, with equal-length wings (50 mm) (after Kukalová-Peck 1972). (b) *Protohymen permianus*, in which the hindwings are a little shorter than the forewings (after Kukalová-Peck 1974).

the differentiation of the pterothorax was much more distinct, the wing base was narrow, its leading edge was strongly costalized, the venation was substantially reduced, there were no transverse veins, and the venation pattern was very consistent. Outwardly they resembled the Megasecoptera, from which they probably inherited petiolate wings and a special mode of wing costalization in which there was a concentration of longitudinal veins parallel to the leading edge (Fig. 6.3).

In addition to the extrinsic reason for the departure from initial synchronous wing pairs working described above, there was also an intrinsic reason. The asymmetry of the meso- and metathorax was presumably mainly determined by the construction of the pterothorax. Typically, the longitudinal dorsal muscles of the mesothorax ($t14$) are attached to the posterior phragma of the mesothorax from the anterior, while the analogous muscles of the metathorax are attached from the posterior. If both pairs of wings are depressed synchronously, i.e. they are moved in phase, then, the muscles of both segments must contract simultaneously, with phragma II (the middle phragma) serving as a support for the muscles of both segments. The contraction of these muscles displaces the anterior phragma backwards, and the posterior phragma (phragma III) forwards. The two segments of the pterothorax appear to be unequally positioned, so that when the muscle $t14$ contracts in the mesothorax, the anterior part of the tergum moves backwards; whereas in the metathorax, the posterior part of the tergum moves forward when the analogous muscles are simultaneously contracted. This asymmetry must have meant that sooner or later the original synchrony in wing pairs depression was replaced by different phase relations. As a result, four different types of specialization of the action of the longitudinal dorsal flight muscles have arisen.

In 'hind-motor' insects (the Blattodea, Orthoptera, and others) the hindwings have become the principal flight organ, so that the stroke amplitude of the forewings, and hence the loading on the longitudinal dorsal muscles of the mesothorax have diminished. Due to this, the contraction of the metathoracic $t14$ muscles shifts the middle phragma backwards. In mayflies, on the other hand, the longitudinal dorsal muscles of the metathorax have become considerably reduced, as the hindwings have become completely subordinate to the forewings. In this case, the contraction of the mesothoracic *t14* muscles shifts the middle phragma forwards. In functionally four-winged insects, both pairs of wings retained their autonomy of movement and were depressed synchronously, so that either both segments of the thorax became equally developed, or there was some domination of the mesothorax. Here, the asymmetry of meso-

and metathorax led to the phase-shifted end of downstroke of the fore- and hindwings and finally resulted in the appearance of functionally two-winged flight. In dragonflies, which represent the fourth line of specialization of the longitudinal dorsal muscles, powerful development of the direct wing depressors has provided autonomy of wing function to both segments, and hence to both sides of the body. The underlying principles of the functioning of these muscles have changed so radically in the Odonata that the longitudinal dorsal muscles have become reduced to two slender bundles.

6.2 The eosynchronous flight of mayflies

Stroke amplitude is very high in mayflies, and the longitudinal dorsal musculature plays the main part in wing depression. Presumably, the aerodynamically profitable increase in stroke amplitude was linked to an increased role of *t14* in wing motion. In this situation, the problem of optimizing the interaction between the wing pairs must have been of overwhelming importance. When the fore- and hindwings work synchronously, the wings of one side can be either spread apart or pulled together during the downstroke, in the latter case with the forewing partly overlapping the hindwing, facilitating coordinated depression of the wings. The wings are only spread apart during the downstroke in the zygopterous Odonata and in some functionally four-winged insects. Since the wings of the meso-and metathorax of ancient insects were not coupled, and were depressed independently, it is quite logical to suppose that the forewing did overlap the hindwing. The increased role of the dorsal longitudinal musculature in wing movement facilitated the synchronous depression of the wings, as these muscles have the same site of insertion (the posterior phragma of the mesothorax) in the meso- and metathoracic segments, which is why the muscles necessarily contract concurrently. For these reasons, phase-shifted wing pairs motion in the Ephemeropteroidea (both the Permoplectoptera and the Ephemeroptera) analogous to that in the Odonata and some Megasecoptera appears

to have been impossible. In mayflies, movement of the mesonotum is transferred to the wing through the posterior notal wing process and the trailing edge of the wing (and from there on to the leading edge of the hindwing), which is why these insects have retained the intimate position of the wings for the downstroke as well as their original stroke synchronicity. Hence the reinforcement of the dorsal longitudinal musculature of the mesothorax enabled mayflies to retain their original synchronous wing pair movements. This type of wing kinematics may be called 'eosynchronous'.

The main features of the structure of tergum of the wing-bearing segment in these insects are the unclosed branches of the scutoscutellar suture and the absence of the parts of these branches which lie posteriorly and externally from the main hinge (*h1*) of the tergum (see Fig. 1.7, p.12). The scutoscutellar and recurrent scutoscutellar sutures are equally well developed. The hinge at the intersection point of these sutures provides the scutellar arm with great mobility. Wing steering is accomplished via the posterior notal wing process so that contraction of the dorsal longitudinal muscles causes the scutellar arms to rotate in their hinges, thus elevating the posterior notal wing processes. Furthermore, the strongly extended anterior part of the notum and long prescutum should be pointed out as characteristic features of the structure of the mayfly mesotergum. Similarly extended anterior part of the notum is observed in the Hymenoptera Aculeata, the Diptera, and the Bittacidae in the Mecoptera. Union of the muscles in morphofunctional bundles is absent in mayflies; all of the dorsoventral muscles are attached separately to the notum (Fig. 6.4). The area of insertion of the upper ends of the longitudinal dorsal muscles spreads a long way backwards, which determines the characteristic form of the parallel parapsidal sutures. The lateral parapsidal sutures are well developed. Thus contraction of the longitudinal dorsal muscles in mayflies brings about backward movement of the scutum, so that its posterior part, which is separated from the anterior part by the oblique sutures and bears the scutellar arms, rotates and elevates the posterior wing processes, which in turn elevate over both the anterior and the pleural wing processes. In consequence, the

wings not only go down but also rotate about their longitudinal axes, with their leading edges downwards, i.e. pronate.

The complete set of mayfly wing muscles is shown in Figure 6.4. The presence of a large number of tergosternal muscles is characteristic of their wing apparatus, as is the powerful development of the muscles of the subalar complex and weak basalar muscles. The latter muscles exert slight additional control of wing movement, this function being more pronounced in relatively advanced species (Brodsky 1974).

During flight, mayfly wings move along a steep trajectory, with a large amplitude and low frequency. The hindwings have lost their autonomy, and are moved by the forewings, resulting in partial reduction of the hindwings. Their rate of reduction depends on their role in gliding and parachuting. To retain its gliding

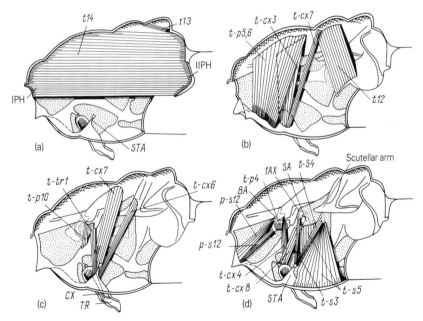

Fig. 6.4 Wing muscles of the mesothorax of the mayfly *Ephemera vulgata*. (a) Sagittal cross-section through the mesothorax, showing a huge *t14* muscle; (b) medial muscles removed, revealing the dorsoventral and dorsal oblique muscles; (c) the central group of muscles (note that the upper end of muscle *t–cx6* is inserted on the scutellar arm, thus controlling its movement during wingstrokes); (d) direct flight muscles. The muscle set of the subalar sclerite is composed, apart from *t–cx8*, of tergosternal muscles. The powerful muscle *t–s5* provides strokes with a larger amplitude when an insect gains height during a mating dance, whereas the slender tonic muscles *t–s3* and *t–s4* lock the extended wings during parachuting.

ability a mayfly must possess small hindwings which form a flap at the trailing edge of the forewing. If the flap is lowered, wing camber, and thereby lift, are enhanced; if it is down on the right wing only, the mayfly will roll to the left, inducing a left turn. As shown by Brodsky and Ivanov (1975), mayflies change the size of the gap between fore- and hindwing, thus altering the airflow over the hindwing. Only small-sized mayfly species such as the Caenidae, some Baetidae, and few others have completely lost their hindwings. When in flapping flight, a flying mayfly leaves a vortex wake in the surrounding air whose sagittal section forms a false straight vortex street in the imago and a straight vortex street in the subimago.

Trivial flight does not play a very significant part in the life of the mayfly, and is unsteady. It is often of short duration: the flying insect lands as soon as possible. The flight path is erratic due to frequent changes in height, and the longitudinal body axis is not always turned in the direction of flight. Upwards spiralling flight is typical, especially in small mayfly species. Their flight velocity is usually low. Frequent changes in flight height, and yawing movements may be caused by uneven working of the wings. But the most outstanding feature of mayfly flight is their elegant mating dances along river and lake banks. The mass flight of millions of mayflies is no less impressive. Mayflies begin their nuptial flights immediately after shedding the larval cuticule, and since they have very little time, they emerge from the water synchronously and in such numbers that the water seems to be 'boiling'. A seething column of insects rises over the lake or river. People who have watched the mass flight of mayflies describe it as a phenomenon on a grand scale that makes an unforgettable impression.

6.3 The bi-motor flight of the Odonata

Notwithstanding the great similarity between the hindwings of *Eugeropteron* and *Geropteron*, they do differ. First, the hindwings of *Geropteron* are 1.5 times as long as those of *Eugeropteron*. Second, the torsional hinge of *Geropteron* has a simpler structure, which implies that it had a lower stroke amplitude than *Eugeropteron*. There must presumably have been a tendency for the wings to length and for the stroke amplitude to decrease in the Geroptera at some stage. Thus, via processes analogous to those which in another evolutionary branch led to the formation of the Palaeodictyopteroidea, a branch diverged from the Geroptera, giving rise to the giant dragonflies, the Meganeurida. The wings of their earliest representative *Erasipteron larischi* (Fig. 6.2a) combine the characters of the Geroptera and the Meganeurida, as was demonstrated by Riek and Kukalová-Peck (1984). Adaptations for gliding undoubtedly played an important role in the development of the Meganeurida, although the wing of *Erasipteron larischi* is still less adapted for gliding flight when compared with the hindwings of anisopterous Odonata. Some characters of the meganeuridan wing organization, such as the lack of nodus or pterostigma, separate the giant Palaeozoic dragonflies from the true dragonflies. True dragonflies also passed through a stage of adaptation for gliding flight, but nowadays their wings are highly specialized to allow strong deformation due to the presence of numerous torsional hinges, such as the nodus and arculus, among others. The perfect flight of dragonflies is the peak in the evolution of four-winged flight.

6.3.1 Structure of the wing apparatus

The structure and mechanics of the pterothorax of dragonflies have been described repeatedly (Tannert 1958; Neville 1960; Hatch 1966; Deshpande 1984, etc.), but views on the homologies of the tergal elements are still quite contradictory. Matsuda's opinion (1981) of the origin of the wing apparatus of dragonflies, independent of all other insects, exemplifies an extreme version of existing contradictions. As has already been mentioned, dragonfly wings move by contraction of the direct flight muscles: this is associated with poor development of the terga of the meso- and metathoracic segments. To study their structure it is useful to examine the organization of the dorsum of some extinct dragonflies belonging to the genus *Arctotypus*, which were members of the earliest and most primitive suborder of the Meganeurida (Pritykina 1989). The Meganeurida had small terga (compared with body size) as do present-day Odonata. At the same time, a number of tergal structural characters bring them close to contemporary Ephemeroptera (Fig. 6.5). First, both the Meganeurida and Ephemeroptera are characterized by a complex and well developed acrotergite and an elongate anterior part of the notum, resulting in long prescuta and anterolateral scutal sulci in the form of long deep grooves. In addition, the shape of the recurrent scutoscutellar suture is similar in mayflies and meganeurids. Dents at the site of attachment of the dorsal oblique muscle and a short sharp ridge on the posterior part of the scutum, which delimits the sites of attachment of the muscle *t13*, emphasize the similar organization of the scutoscutellar suture. However, the meganeurids lacked parapsidal and lateral parapsidal sutures.

The parapsidal and lateral parapsidal sutures are also not developed in contemporary dragonflies (Fig. 6.6). They resemble the meganeurids in such features as a complex acrotergite, long,

peculiar shaped prescuta, deep anterolateral scutal grooves, the presence of a hinge (*h1*) at the origin of the oblique sutures, and presumably in the similar structure of the posterior part of the notum. The most characteristic feature of the organization of the tergum of modern Odonata is the reduction of the posterior part of the notum, and the excessive development of the anterior part, with the assistance of which the wing is steered, via the humeral and radial axillary plates. The humeral plate comprises the humeral plate itself, tightly joined with the basalar sclerite (Fig. 6.6b); the radial axillary plate was formed as a result of the complete or partial fusion of the second axillary sclerite with the third axillary and also with the basiradiale and the bases of the medial and cubital veins (Fig. 6.6b). The transscutal sulcus, which caused the reduction of the scutoscutellar suture, lies in front of the anterior notal wing processes. Dorsal oblique muscles are absent. The recurrent scutoscutellar suture is clearly distinct, as are the hinges, from which oblique sutures arise which expand laterally (Fig. 6.6a). It should be pointed out that analogous expansion of the oblique sutures and their transformation into fissures can be seen in some advanced mayflies (Brodsky 1974).

Only the *t14* indirect flight muscle of recent

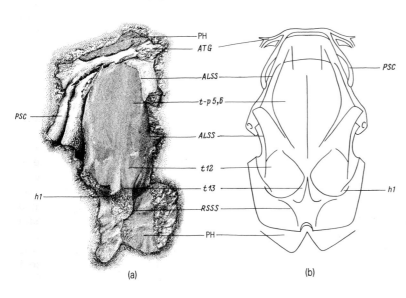

(a) (b)

Fig. 6.5 Terga of an ancient giant dragonfly and a present-day mayfly. (a) Dorsal view of the metanotum of the meganeurid dragonfly *Arctotypus* sp. (b) Dorsal view of the mesonotum of the mayfly *Ephemera vulgata*.

Odonata can be reliably homologized with those of other insects (Fig. 6.6b). The dorsal longitudinal muscle is inserted anteriorly on a phragma which results from the invagination of the antecostal suture. The prescutum bears a large heavily sclerotized part laterally which distally articulates with the humeral plate and supports a muscle which Pfau (1986) designated the 'first dorsoventral muscle' proximally (Fig. 6.6a). It should also be pointed out that the Odonate wing apparatus lacks bifunctional muscles. Only one leg muscle (*t–cx4*) participates in wing movement. This feature enables dragonflies to move their legs in flight freely, which is necessary for seizing prey. The musculature sets of the meso- and metathoracic segments of zygopterous Odonata are quite similar to each other, whereas in the anisopterous Odonata one of the mesothoracic tergopleural muscles (*t-p11*) is attached to the edge of the scutum, although not to the tergal apophysis. It has been suggested (Deshpande 1984) that this muscle has retained its primitive condition in the mesothorax.

6.3.2 Wing apparatus function

Casual observation of the free flight of any member of the order Odonata suggests that the movements of the fore and hind pairs of wings are unsynchronized. However, careful study of the wing kinematics of members of several families reveals quite the opposite.

A phase difference of one half of a stroke cycle between the movements of the fore- and hindwings was observed during the hovering flight of *Aeschna juncea* (see Fig. 4.5, p.72) and during the tethered flight of *Sympetrum flaveolum* and *S. vulgatum* (Brodsky 1988). In the anisopterous Odonata the phase shift is commonly about a quarter of a stroke cycle, with either wing pair leading. In either situation, the stroke amplitude of one of the pairs (usually the fore pair) is greater than that of the other. In the damselfly *Coenagrion hastulatum* there is a phase shift of a quarter of a stroke cycle, with noticeable domination by the hindwings. In *Lestes sponsa*, on the contrary, the forewing stroke amplitude was greater. Moreover, the fore- and hindwings often move in different planes, the

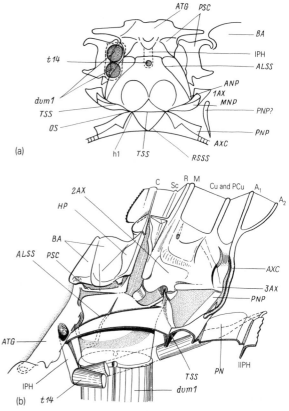

Fig. 6.6 Structure of the dragonfly pterothorax. (a) Dorsal view of the mesonotum of the dragonfly *Libellula quadrimaculata*. (b) Lateral view of the inside of the mesothorax of an aeschnid dragonfly, showing the sites of attachment of the muscle *t14* (after Pfau 1986, with altered labels).

forewings moving above the horizontal plane and the hindwings below it, or vice versa, and the wing stroke planes are inclined differently relative to the longitudinal body axis. It is common for the wings of each side to move differently when in tethered flight; one of the wings may even stop moving for some time. However, according to Alexander (1984), some anisopterous Odonata can produce several synchronous flaps with both pairs of wings when taking off or hovering.

The phase shift between the wing pairs is most

often one quarter of a stroke cycle, with the hind-wings leading, especially during fast forward flight. This phase relationship is strictly adhered to in the downstroke, whereas there is consider-able variation during the upstroke. Azuma and Watanabe (1988) showed that in the dragonfly *Anax parthenope* the phase angle between the hind- and forewings during the upstroke varied from one third to a sixth between individuals, but did not deviate much from a quarter during the downstroke. A phase shift of one quarter of the stroke cycle means that at the ends of the up-and downstrokes the forewings decelerate as the hindwings accelerate (Fig. 6.7).

Mechanisms with an inertial recuperative drive whose mass units are connected by a transmit-ting device which provides strong linkage and kinetic energy exchange show similar timing in their acceleration and deceleration to that described above. This type of oscillating sys-tem (an inertial kinematic oscillator) provides a 10–100 times decrease in the consumption of power and energy during the oscillation of the executive units. The graph showing the move-ment of the fore- and hindwings of the dragonfly (Fig. 6.7) lends support to the belief that its wings are connected by a transmitting device during the acceleration–deceleration phases thus forming an inertial kinematic oscillator which provides recuperative kinetic energy exchange between the wings. To prove the existence of this mechanism, it is necessary to discover whether there is a mechanical link between the fore- and hindwing pairs.

A simple experiment by Brodsky and Judovsky (in press) with live immobilized dragonflies *Calopteryx splendens* (both males and females) showed that when the forewings were forcibly depressed from the vertical into a position in which the angle between the longitudinal wing axis and the vertical was 60°, the hindwings remained immobile; however when the fore-wings were depressed further, the hindwings started to move. The hindwings moved to 20° from the vertical when the forewings were in the position which corresponds to the lowest point of the wing path during flight. Analogous results were obtained by Alexander (1984). He remarked that depression of the bases of the

hindwings caused elevation of the forewings, and that depression of the bases of the forewings affected the position of the hind ones. Moreover, he emphasized that these movements only took place when the wings approached the extreme points of the stroke. This data proves that a mechanical link between the wing pairs does exist, but that it only starts to work when the angle between the forewings and the vertical reaches 60°, and then continues untill the end of the stroke. It is at this stage that deceleration of the forewings and acceleration of the hindwings take place during flight (Fig. 6.7), suggesting that there might be recuperative energy exchange between the wing pairs. An analogous energy exchange may also exist at the upper point of the stroke, when the forewings decelerate and the hindwings accelerate. However, the experiments with live immobilized insects discussed above do not provide any evidence on this point.

Let us consider what form this mechanical linkage may take. It is likely to consist of skeletal elements which are distorted by the deceleration of the forewings, as well as by the acceleration of the hindwings. In the typical insect pterothorax this type of linkage can only be provided by phragma II (the posterior phragma of the meso-thorax, which is the anterior phragma of the

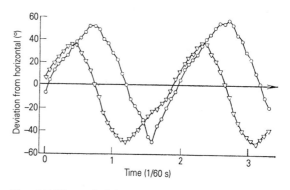

Fig. 6.7 Phase shift between movements of the fore- and hindwings during tethered flight of dragonfly *Ladona exusta* (after Chadwick 1940). The ordinate is the angle between the longitudinal axis of the wing and the horizontal, and the abscissa is time. Open circles, forewings; triangles, hindwings.

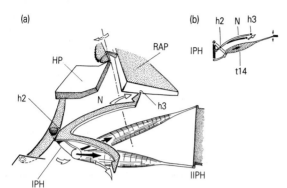

Fig. 6.8 The arrangement of the dragonfly meso-notum during contraction of the muscle *t14* (after Pfau 1986). The shift in position of *IPH*, and of the notal margin near *h3*, as *t14* contracts, is shown by a light arrow. *RAP* is the radial axillary plate. (a) General view from inside. (b) The position of *IPH* and *N* when shifted is shown in light outlines.

metathorax), the attachment point of the dorsal longitudinal muscles of both segments. In the Odonata mesothorax the anterior ends of this muscle are inserted on the inferior margin of a narrow, two-lobed phragma I (Fig. 6.8a). Its posterior ends diverge broadly and are attached to the antecostal suture of the metatergum which gives support to this muscle, in the same way as in the wing apparatus of other insects. When muscle *t14* contracts, it pulls the inferior margin of the phragma I backwards, so that the notum which is connected to the phragma by a mobile articulation, *h2*, curves and elevates (Fig. 6.8b). The beak-shaped notal margin, which corresponds to the first axillary sclerite of other insects, articulates with the radial axillary plate (*RAP*) via the hinge *h3* (Fig. 6.8a). The latter transmits the movement of the notal margin onto the radial axillary plate causing it to rotate around the transverse axis (Fig. 6.8a) with its posterior edge downwards. In dragonflies, numerous powerful subalar muscles are attached to the radial axillary plate. Action of the posterior subalar muscles (Fig. 5.6b, p.91) is analogous to that caused by the contraction of the longitudinal dorsal muscles, but is considerably stronger.

Moreover, the *t14* muscles are so weak that the movements of the sclerites described above are only theoretically possible; in fact, depression and turning of the radial axillary plate is carried out exclusively by the powerful subalar muscles. The greatly reduced mesothoracic longitudinal dorsal muscles presumably function as a mechanical link between the meso- and metathorax.

It is known that the wing musculature is served by fast, slow, and damping axons. Their firing in various combinations may cause tetanic as well as tonic contractions. In the latter case the muscle loses elasticity, forming a strong link between the inferior end of phragma I and phragma II. This link may transmit kinetic energy from the forewings to the hindwings as phragma I is moved forward (Fig. 6.8b, in the opposite direction to the light arrow), and it breaks off as phragma I moves back.

As the forewings decelerate, the wing base sclerites move in the opposite direction. As the basalar and subalar muscles relax, the posterior edge of the radial axillary plate turns upwards, while simultaneously depressing its proximal edge because of elasticity in the articulation area *h3*. The notum depresses and shifts the phragma I forward (Fig. 6.8b, in the opposite direction to the arrow). As phragma I turns through an angle, the *t14* muscle, being tonically contracted, pulls the phragma II downwards. The hindwings move upwards.

The elasticity of the *h3* articulation is controlled by the fulcroalar muscle (*t–p14*) which is attached to the posterior edge of the radial axillary plate. Newman and Wootton (1988) studied the action of this short powerful muscle and showed that its contraction causes deformation of the radial axillary plate so that its anterior and posterior walls draw together, thus increasing tension in the hinge *h3* area. When the forewings, having passed the lowest point of the stroke, begin the upstroke, the *t14* muscles relax and the mechanical link between the fore- and hindwings terminates.

Thus the mechanical link which brings about depression of the metanotum and elevation of the hindwings engages at the point when the forewings pass below the horizontal. The kinetic

energy of the decelerating forewings is transmitted to the elevating hindwings through the
system of sclerites (the radial axillary plate, the
notum, phragma I, and phragma II) and the contracted longitudinal muscles of the mesothorax.

Variable phase relationships in the movement
of the wing pairs are not the only unique feature
in the design of the wing apparatus of Odonata.
They also have a special mode of wing deformation during the strokes. Numerous hinges
cause rapid local deflections of different parts
of the wings throughout the strokes. A strong
screw-like torsion at the bases of the wings
occurs in the zygopterous Odonata, while in
contrast, the anisopterous Odonata are characterized by a wave-shaped deformation which
passes along the trailing edge of the wings.
When an anisopterous dragonfly hovers (see
Fig. 4.5, p. 72), a deformation wave passes
through the wing from tip to base. As a result,
the vortex wake produced by a flying dragonfly
has an unusual shape.

During a second type of hovering, dragonfly
wing pairs move in antiphase: the forewings
depress, the hindwings elevate, and vice versa.
This type of vortex wake formation was studied
by Brodsky (1988) in the tethered flight of the
dragonfly *Sympetrum flaveolum*. The most striking aspect of dragonfly flight is the active role
of each of its four wings in vortex formation.
The most intense vortex formation occurs at
the trailing edge of the hindwing. During the
downstroke, a strong air current flows over the
upper wing surface to form a starting (dorsal)
vortex, analogous to that of other insects with a
low wingbeat frequency (see Chapter 3). When
the hindwing reaches the lowest point of the
trajectory, the current flowing down the wing
turns into a stopping (ventral) vortex (Fig. 6.9a).
By the end of the downstroke, the dorsal vortex
migrates downwards and slightly backwards.
While the wing supinates at the bottom of
the stroke, a deformation wave passes through
it from tip to base, followed by an intensive
shedding of the vortex sheet. As a result of
this movement, a small diameter vortex ring is
created, that is, the stopping vortex completes
the missing part of the ring (Fig. 6.9b). As
the hindwing rises the ring is shed and shifts

downwards and backwards. Small vortex rings,
shed from the right and left wings during the
strokes form two vortex streets which point
downwards, backwards, and slightly away from
the insect's body.

Vortex rings are produced in the same way by
the forewing strokes. When the forewings are at
the bottom of the stroke, small vortex rings shed
from the wings are thrown aside. After several
strokes, two chains of small vortex rings which
are oriented forwards, downwards and away
from the insect's body are formed. Thus four
differently oriented vortex tracks are produced

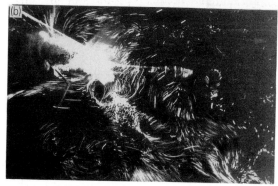

Fig. 6.9 Some essential features of vortex formation in tethered flight of the dragonfly *Sympetrum
flaveolum*. Side view, head to the left, with a
vertical light plane passing through the bases of
the left-hand wings. Stroke parameters: $n = 30$ Hz,
$A_{fw} = 70°$, $A_{hw} = 65°$, $b_{fw} = 50°$, $b_{hw} = 40°$, $B = 0°$.
Phases of the hindwing stroke: (a) the end of the
stroke, and (b) the beginning of the upstroke.

by a hovering dragonfly. If these four tracks were superimposed, the resulting picture of a vortex wake would correspond to the reconstruction suggested by the four Japanese investigators (Azuma *et al.* 1985) for the dragonfly *Sympetrum frequens* in slow rising flight.

As can be seen from the data mentioned above, in dragonflies the leading role in vortex formation is played not by the strokes themselves, as in other slow-flapping insects, but by a comparatively slow and strong tip-to-base deformation wave which passes along the wing at the lowest point of the stroke. When the dragonfly is doing the second type of hovering, the wing stroke amplitude is small, and the deformation wave is rather strong. When the dragonfly is flying forwards, one would expect the vortex wake to consist both of large diameter rings produced by flapping movements of the wings and of a number of small diameter rings produced by the strong deformation waves passing through the wings.

All the above applies only to the anisopterous Odonata. How the vortex rings used by zygopterous Odonata are generated is not yet clear: on the one hand, each of the wings works independently, on the other, wave-type deformations do not normally occur.

6.3.3 Dragonfly flight

Long before the era of the first dinosaurs, giant dragonflies, some of them with a wingspan of almost a metre, the largest flying insects of all times, were gliding over the tropical lagoons. The flight pattern of modern Odonata has changed considerably, and also varies considerably between dragonflies, damselflies, and Calopterygidae. The flight pattern of zygopterous Odonata is characterized by comparatively low velocity flight in any direction without turning of the body. It is when hunting in grassy vegetation that the zygopterous Odonata reveal their flight abilities the best; the ability to alter flight direction instantaneously, to move sideways or at any other angle relative to the longitudinal body axis. The first type of hovering flight is customary for them. Climbing is accomplished in one instantaneous leap; the damselfly is like

a compass arrow in the way it keeps its horizontal position.

The Calopterygidae are much slower, poorer fliers, and they only flutter; although unlike other dragonflies, males dance in front of females. These dragonflies are also capable of both short diving-gliding flights with their wings lifted and folded over the back and compound manoeuvres involving a sequence of simple turns. According to Rüppell (1985), *Calopteryx splendens* displays two fundamentally different flight modes, one with approximately synchronous fore- and hindwing strokes and another with approximately alternating fore- and hindwing strokes. Switching from one mode to the other in midflight is possible and can be observed quite frequently. The male banded agrion is able to fly with a very low wingstroke frequency when using aerodynamically effective synchronous flight. This probably increases the apparent surface of the blue wings and, consequently, also their signal effect.

The anisopterous Odonata have a quite different flight pattern. They do the second type of hovering flight. The most striking feature of their flight is their ability to turn rapidly almost on the spot: a flying dragonfly stops abruptly, and turns in the right direction while hovering, and then continues flying forwards. The turn may be up to 90° or more. They climb fast, without turning the body upwards. The high turning speed is accompanied by an amazingly complicated pattern of body position changes during flight. Ryazanova (1966) observed dragonflies of the genus *Sympetrum* performing such complex routines as rotating about the longitudinal body axis with concurrent forward movement, and in *Libellula quadrimaculata*, change to the opposite flight direction may be accomplished by a somersault and a brief flight upside down. Nearly all small dragonflies of the family Libellulidae are capable of brief flights upside down and also of rotation about the longitudinal body axis.

Thus, to summarize all that has been said about dragonflies above, a unique flight mechanism has developed in the Odonata in the course of evolution, based on the functional autonomy of the meso- and metathoracic wings. This system of wing motion cannot provide rapid flapping, but has the advantage of allowing each of the four wings to function independantly, which enables the Odonata to perform complex manoeuvres in flight. Their flight can be considered truly bi-motor.

6.4 Conclusion

A low wingbeat frequency (20–30 Hz) and frequent transitions to gliding flight were usual in early forms of flight. The phase relationships of the two pairs of wings were not stabilized, varying from symphase in the Palaeodictyoptera to antiphase in some Megasecoptera. Synchronous movement of the wing pairs was common in all Palaeodictyoptera and Diaphanopteroidea, and still occurs in present-day Ephemeroptera. As a result of the synchronous movement of the wing pairs, dipterigy (two-wingedness) has developed in some Ephemeroptera and in the Permothemistida and Eukulojidae. Synchronous functioning of the two wing pairs, sometimes observed in anisopterous Odonata, is presumably a curious 'echo' of wing functioning in their remote ancestors. The different phase relationships in wing pairs movement which occurred in some Megasecoptera are now typical of all Odonata. Petiolate wings developed in the Megasecoptera and in zygopterous Odonata, probably associated with an increase in the ability to vary the stroke plane relative to the body axis. On the basis of gradual supination, which is usual in mayflies and their ancestors, new types of wing deformation appeared: the screw-like torsion of a petiolate wing and a deformational tip-to-base wave provided by numerous hinges in the wings of anisopterous Odonata. The posterior notal wing process is predominant for directing the wings in the Ephemeroptera and Odonata, and its development was in turn accompanied by progressive development of the subalar muscles. In some flight modes the wing apparatus of the Odonata functions as an inertial kinematic oscillator, which provides a recuperative kinetic energy exchange between the wing pairs.

7

Flight based on hindwings

If the test of nobility is antiquity of family, then the cockroach that hides behind the kitchen sink is the true aristocrat.

Sutherland

At the roots of the evolutionary line which leads to the present-day Neoptera there was an extensive order with hazy borders, the Protoptera, which is only known from wing fossils. Various members of the lower Para- and Oligoneoptera, especially the orders Hypoperlida and Blattinopseida, as well as the Polyneoptera can be derived from them via smooth transitions. The Polyneoptera evolved a more concealed lifestyle in litter, under bark, and in rotting wood and especially in accumulations of coarse plant remains—under fallen trees, twigs, and branches and in 'macrolitter' (Rasnitsyn 1980a). Transition to a concealed lifestyle could have been caused by the appearance of aerial predators—dragonflies, which seem to have appeared a little before the Polyneoptera. The most compact body form is provided if the wings are folded flat over the abdomen. With this manner of folding, the forewings acquired a protective function, they became hardened, and the flight function was concentrated in the hindwings. The hindwings expanded at the base to form a canopy during flight (Antonova et al. 1980), flexing in a complex manner for wing folding.

7.1 The quasi-synchronous flight of stoneflies

Stoneflies appear relatively late in the palaeontological records—the most ancient members of the order are known from the end of the lower Permian. It is now indisputable that the order originated from a group of terrestrial Grylloblattida, but which ancestral group the stoneflies came from is unknown (Sinichenkova 1987). There are only a few, wingless, present-day representatives of the Grylloblattida. We may therefore suppose that the stoneflies are the most primitive members of the contemporary winged Polyneoptera. They have retained a lot of primitive structural features, such as thysanuran-like nymphs, the meso- and metathorax movable relative to each other, the absence of a distal medial plate in the axillary apparatus, an equal stroke amplitude by fore- and hindwings, the same yawing flight as in mayflies, and so on. All this makes stoneflies especially interesting.

7.1.1 Structure of the wing apparatus

The tergal parts of the meso- and metathorax have the same structure, only differing in size. The characteristic feature of stonefly tergal structure is the absence or poor development of the scutoscutellar suture and powerful development of the recurrent scutoscutellar suture. The large area of attachment of the upper end of muscle $t12$, located medially, is not delimited by a suture posteriorly (see Fig. 1.8a, p.13). Only in some advanced species, like *Kamimuria luteicauda*, is there a very poorly pronounced suture at this site (Fig. 1.8b). The right and left sutures meet medially, dividing off a small part of the scutum into a shield. However, unlike mayflies, stonefly muscles are not attached to the branches of the scutoscutellar suture from posteriad, so there is no true scutellum. Muscle *t20*, if present, is attached outside the

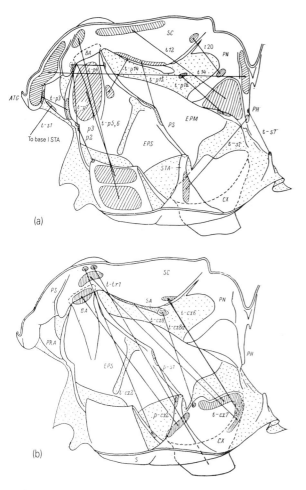

(a)

(b)

Fig. 7.1 The wing muscles of stoneflies. Sagittal cross-section of the wing-bearing segment, inner view. (a) and (b) show different muscle groups. The sites of muscles attachment are hatched. Compared with mayflies (Fig. 6.4), stoneflies have powerful basalar muscles and a large number of bifunctional ones, mainly tergocoxal and pleurocoxal. The tergosternal muscles are almost absent.

scutoscutellar suture. Among the specialized features we should mention the union of several tergocoxal muscles into a morphofunctional bundle that is attached to the notum opposite the anterior wing process together with muscle *t–p5*, *6* (Figs 1.8 and 7.1). The bifunctional muscle *t–cx6* is developed relatively strongly. It should be pointed out that the scutoscutellar suture, if present, does not affect notum deformation when the dorsal longitudinal muscles contract. Some deformability of the notum is provided by the lateral parapsidal sutures which separate the insertion sites of the tergocoxal muscles and *t12* (Fig. 1.8). In comparatively primitive stoneflies these sutures are especially well developed and look like scutoscutellar sutures, and they have been mistaken for the latter in numerous works on plecopteran taxonomy.

The pleura are clearly divided by the pleural suture into a wide episternum and a somewhat reduced epimeron (Fig. 7.1). In some species the epimera are strongly linked with the lateral postnotal processes forming postalar bridges, while in other species a membranous field lies between these. The prealar arms of the scutum are not fused with the episterna.

The articulation areas of stonefly wings are characterized by an extremely archaic structure. Their axillary apparatus differs from the generalized one (see Fig. 1.10, p.16) by the absence of a distal medial plate and a different position of the humeral fold. The basiradiale is partly fused with the second axillary sclerite and forms a canopy-shaped structure which flexes along the humeral fold (Fig. 7.2) when the wing is folded backwards.

Stoneflies have two wing pairs; the hindwings are a bit shorter and broadened at their bases (see Fig. 1.19b, p.29). Both pairs are folded flat over the abdomen. The forewings completely overlap the hindwings, protecting the insect from mechanical damage. Stonefly hindwing venation can be quite easily derived from that of the Protoptera hindwing; the most substantial differences are in the structure of the anal area. In the Protoptera the hindwing anal area was already broadened (see Fig. 1.19a), but it was presumably not deflected at wing folding. In stoneflies accessory broadening of the

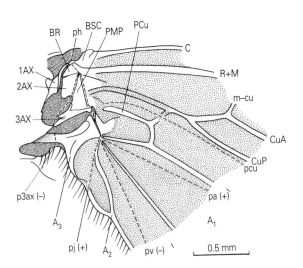

Fig. 7.2 Hindwing base of the stonefly *Isocapnia guentheri.*

vannus took place, mainly due to A_1 forking. From that moment the anal and vannal folds caused deflection of the vannus during wing folding, complementing the jugal fold (Fig. 1.19b). In the case of the secondary narrowing of the hindwings in Capniidae *PCu* is reduced, and all three anal veins are retained (Fig. 7.2).

7.1.2 Wing apparatus function

In stoneflies both wing pairs move with approximately equal amplitude and the phase shift is about a twentieth of the stroke cycle, but both pairs of wings reach the stroke extremes simultaneously. The trajectories drawn by the tips of the fore- and hindwings are of a complicated form and are differently inclined to the horizontal (Fig. 7.3a and Fig. 2.7b and c, p.39). These characters, as well as a slight asynchrony in the movement of the wing pairs seem to result from the first timid step made by stoneflies or their ancestors to break the original synchrony of the wing pairs motion.

The deformability of the wings in flight has changed more than has wing kinematics. Deformability of both the fore- and hindwings has increased considerably between the first

winged insects and stoneflies. In Fig. 7.4 a complete stroke cycle of *Isogenus nubecula* is represented by a sequence of photographs. Deformation of the wings is less marked at the downstroke. When the hindwings are at their highest point, their dorsal surfaces are

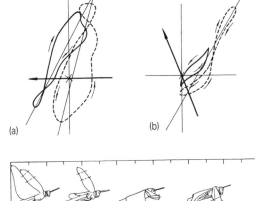

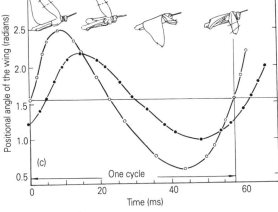

Fig. 7.3 Wing kinematics of stoneflies and hind-motor insects. (a and b) Trajectories described by the wing tip of (a) the stonefly *Isogenus nubecula* and (b) the cockroach *Periplaneta americana* in tethered flight. The arrow indicates the direction of insect head, the cross is centred at the base of the left forewing, and the hindwing tip path is shown by a broken line. (c) Standard wing stroke of locust *Schistocerca gregaria.* Typical wing positions are shown above the curves. Filled circles correspond to the forewing, open circles to the hindwing. The amplitude of the hindwing exceeds that of the forewing. Time begins when the hindwings are half-way up (after Weis-Fogh 1956).

clapped together; as the downstroke begins the line of contact between them gradually moves backward (Fig. 7.4,1). At the middle of the downstroke the vanni of the hindwings form a canopy (Fig. 7.4, 2).

The most severe deformation of the wings occurs during supination, at the lowest point of the stroke. The stages in wing supination are shown in Fig. 7.5. As it approaches the lowest point of the stroke, the forewing decelerates (Fig. 7.5,1), twists at its base (Fig. 7.5,2), and then its tip deflects downward (Fig. 7.5, 3). The line of deflection starts at the leading edge of the wing

from the apex of *Sc*, crosses the wing plane obliquely, and reaches the trailing edge near the end of the cubital fold. A sharp flexion line does not occur because the wing tip deflection is smooth; it starts slowly, but then accelerates and comes to a halt so rapidly that the posterior part of the deflected area reaches its extreme position by means of a flick (Fig. 7.4, 3–5). At this stage the wing is strongly twisted at its base and the deflected part of the wing appears to be turned forward and upward (Fig. 7.5, 4). As the wing starts to move upward, the deflected part stretches out abruptly (Fig. 7.5,5). When

Fig. 7.4 Deformation of the wings during one stroke cycle of the stonefly *Isogenus nubecula*, as seen from a series of photographs taken obliquely from above. Stroke parameters: $n = 33$ Hz, $A_{fw} = 125°$, $A_{hw} = 145°$, $b_{fw} = 60°$, $b_{hw} = 75°$, $B = 0°$.

the deflected part has returned to its original position, the wing appears to be fully supinated (Figs. 7.4, 6 and 7.5, 6). In addition, as the wing is twisted at its base the clavus is strongly curved. At the next instant the clavus rises and turns relative to the cubital fold which, acting as a stiffening rib, shifts the torsional deformation apically (Fig. 7.5, 5 and 6). A short pronounced groove runs from the transverse vein which is crossed by the cubital fold to the trailing edge of the wing (Fig. 7.5, 6). Another deep groove runs along the cubital fold itself to the trailing edge.

The hindwings already begin to supinate at the end of the descending branch of the trajectory. Then the remigia turn their anterior edges upward while the vanni bend longitudinally (Figs 7.4, 3 and 7.5, 2). The hindwing

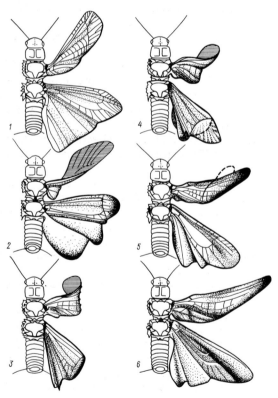

Fig. 7.5 Deformation of the wings of the stonefly *Isogenus nubecula* at the bottom of the stroke: same view as in Fig. 7.4.

tip abruptly deflects downward, forming a sharp flexion line (Figs 7.4, 4 and 7.5, 3). This flexion line lies between the end of the cubital fold and the apex of *Sc*. The grooves are clearly distinct on the deflected part of the remigium, the apical sections of the veins become elevated, and the membrane between them is deflected (Fig. 7.4, 5). At the lower point of the trajectory the vannus is folded repeatedly, the folds overlapping each other (Fig. 7.5, 2, 3). As wing supination progresses, the remigium bends longitudinally along the cubital fold and the fold of the radial sector. The second bend is especially distinct; as it develops, the deflected part of the wing tip stretches out in the same abrupt manner as during deflection (Figs 7.4, 5 and 7.5, 4). Later on, the bend along the radial fold develops.

Thus deformation of the stonefly *Isogenus nubecula* forewings is dominated by screw-like torsion of the surface and elastic oblique wing tip deflection. This type of deformation might be called 'supinational twisting'. An inclined anastomosis line, a shortened subcosta, and a short clavus all have to be present for it to occur. Since all these elements had already evolved in protopteran forewings (see Fig. 1.19a, p.29), it is possible that their wings also underwent supinational twisting. A quite different deformation pattern is usual in stonefly hindwings: the wing tip deflects along the sharp flexion line, and the wing plane bends longitudinally along the cubital fold and that of the radial sector. This type of deformation, which might be considered 'supinational bending', was not usual in protopteran wings (at any rate not in *Zdenekia* hindwings) and associated with the formation of two new folds, the transverse and the radial sector folds.

The giant stonefly *Allonarcys sachalina*, which flaps its wings almost synchronously, at equal large amplitude, is a suitable species in which to study air currents around flying stoneflies. As both wing pairs pronate synchronously at the end of the upstroke, the starting (dorsal) vortex formed from the circulatory flow around the fore- and hindwings builds up over the vanni of the hindwings (Fig. 7.6). At the end of the downstroke the hindwings shift abruptly backwards, so that the vortices shed by the hindwings

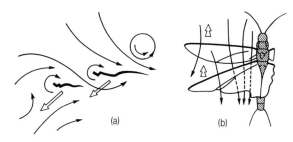

Fig. 7.6 Air currents around the wings of the stonefly *Allonarcys sachalina* at the beginning of the downstroke. Light arrows indicate the direction of wing movement, and thin arrows indicate the directions of air currents. The air currents are shown (a) from the side and (b) from above.

join those from the forewings to form a joint stopping (ventral) vortex. As a result, a common vortex ring is formed during half the stroke cycle. However, the air currents around each wing pair seem to be independent, as can be seen from the airflow through the gap between the fore- and hindwings at different phases of the cycle. If two vortex rings (which unite at the end of each half-cycle) are generated by each half-cycle, one would expect the hindwings to interact with the air more actively as the air currents around them are doubly accelerated. As a result, when the fore- and hindwings pronate and supinate synchronously they generate a common vortex wake, of which the sagittal section shows a false straight vortex street (see Chapter 3). The formation of the false straight street can be explained by the fact that during quasi-synchronous flaps the fore- and hindwings work together, and create vortex rings of different diameters by being strongly supinated. A small diameter ring formed at the downstroke expands as the wings lift, which forms a false straight vortex street.

The vanni of the hindwings of *Allonarcys sachalina* play an important role at the beginning of the downstroke, when the wings of one side work in tandem in what for a moment appears to be in an aerodynamically inappropriate position (Fig. 7.6a); at this stage, the air currents which encounter the ventral surface of the hindwing enter the grooves of the vannus

and are thus drawn off most of the surface (Fig. 7.6b).

7.1.3 The flight pattern of stoneflies

The structure and functioning of the stonefly wing apparatus make it clear that these insects can hardly claim to be expert flyers. Indeed, stoneflies fly reluctantly, and seldom travel a great distance; their trivial flight is reminiscent of mayflies. Some authors (Mertens 1923; Hynes 1974) have described the flight of stoneflies 'heavy and clumsy', emphasizing the poor manoeuvrability of their flight. In general, stoneflies remain quiescent for a long time after moulting into an imago, then fly to various objects such as tree trunks, grass stems, or stones. Migratory flights for a distance of several hundred meters from a river bank occur in some species (Benedetto 1972). It is still not clear whether upstream migrations of females, in compensation for the downstream displacement of nymphs, occur in stoneflies. Mertens (1923) considered that the family Chloroperlidae were the best among stonefly flyers, and observed flight behaviour resembling swarming in which the insects fly rapidly over the water surface, frequently altering their flight direction. The stonefly expert Zhiltzova (personal communication), has also observed swarming, in the families Nemouridae (*Protonemura* sp.) and Perlodidae (*Mesoperlina* sp.), in which the number of insects in the swarm may be so great that it is reminiscent of a mass flight. Swarming can occur both over the water surface and over other landscape features which may resemble an expanse of water, such as a road.

In many species the wing length and the ability to fly are highly variable. Some species always have long-winged females and short-winged males, whereas others may be long-winged or short-winged regardless of sex. Totally wingless individuals do not occur in the large-bodied species. In general, there appears to be a common tendency towards the loss of flight. This tendency is expressed to a greater extent in large-sized species, while small-sized species as a rule fly more swiftly and with greater ease.

7.2 Towards hind-motor flight

A result of the active interaction between the hindwings and the air has been the transformation of the hindwings into the main propulsive organ and the expansion of their bases. The great expansion of the hindwings in the course of evolution led to the appearance of 'hind-motor' insects, in which the stroke amplitude of the forewings became gradually reduced in parallel with the development of the vanni of the hindwings.

Unlike the stoneflies, in which the wing pair synchronicity is relatively disturbed, in hind-motor insects such as cockroaches, mantids, grasshoppers, and their allies, the forewings become gradually excluded from the functioning of the wing apparatus. The wing kinematics of hind-motor insects can easily be derived from those of stoneflies (Fig. 7.3a, p.118) by imagining that the upper part of the forewing trajectory is reduced (Fig. 7.3b and c). The hindwings perform the entire stroke with a large amplitude, both above and below the horizontal plane, while the forewings oscillate only below the horizontal plane. The stroke amplitude of the hindwings is much greater than that of the forewings, and the more stable and swift the flight, the greater the dominance of the hindwing over the forewing stroke amplitude, both in mantids (Brodsky 1985) and locusts (Gewecke 1970).

During the strokes the hindwings constantly beat ahead of the forewings, but as the forewings move with a smaller amplitude, the phase shift between the pairs of wings becomes diminished at extremes of the strokes. The forewings reach the top of the stroke at the same time as the hindwings pass the top of the stroke and begin to pronate (Fig. 7.7, 3). As a result, both pairs pronate nearly simultaneously (Figs 2.3, p.32 and 7.7, 4), so that the hindwings, moving faster, again are ahead the forewings (Figs 2.3, 5 and 7.7, 5), approach the bottom of the stroke, supinate, and go up (Figs 2.3, 9–15 and 7.7, 12–20). The phase shift between the movement of the wing pairs is approximately a tenth of the stroke cycle.

The deformation pattern of the forewings in hind-motor insects is the same as in stoneflies, in which sloping tip deflection (Fig. 2.3, 10–12) and screw-like torsion of the wing blade (Fig. 7.7, 13–19) dominate. The deformation pattern of the hindwings in hind-motor insects is also reminiscent of stoneflies, but is distinguished from the latter by the absence of abrupt wing tip deflection. The vannus has such inertial and aerodynamic resistance that the remigium has enough time to be bent longitudinally along the fold of the radial sector as the costal margin of the wing moves backward and upward at the end of the downstroke (Fig. 7.7, 11–15). This explains why the fold of the radial sector appears to be the main line of supination, and the cubital one is secondary, in wings with a broad anal lobe.

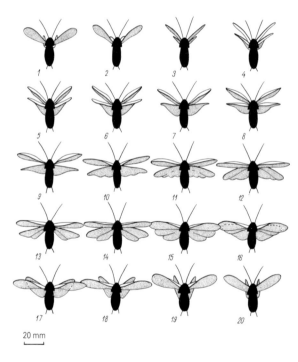

20 mm

Fig. 7.7 Flight of the cockroach *Periplaneta americana* (front view). Consecutive film tracings (1–20) of a single wingbeat. Stroke parameters: $n = 25$ Hz, $A_{fw} = 40°$, $A_{hw} = 110°$, $b_{fw} = 55$, $b_{hw} = 60°$, $B = 67°$. The lower surfaces of the wings are shaded.

7.2.1 The flight system of cockroaches and their allies

The structure of the wing apparatus of hind-motor insects is clearest in cockroaches. Together with mantids and termites they form the super-order Blattopteroidea. That cockroaches, termites, and mantids are closely related is widely accepted and they are quite often united in one order, the Dictyoptera. The late appearance of mantids and termites in the geological record (during the Cretaceous) indicates that they derive directly from cockroaches. The geological history of cockroaches starts from somewhere at the border between the Lower and Middle Carboniferous (Rasnitsyn 1980c). The Spiloblattinidae evolved at the end of the Middle Carboniferous from archaic cockroaches of the family Archimylacrididae, and developed in the direction of flight improvement. This family comprised swift-flying diurnal cockroaches with long membranous wings, which inhabited plants and laid their eggs inside living plant material

with a saw-like ovipositor. Other descendants of the Archimylacrididae became adapted to a concealed lifestyle with weakened flight.

The whole wing apparatus organization of present-day cockroaches shows clear adaptations for a concealed lifestyle. First, the body is flattened dorsoventrally so that the whole structure of the skeleton and thoracic musculature has altered; second, the wings are folded flat; third, the forewings are elytrized, protecting the membranous hindwings from mechanical damage. In addition, movement within the litter requires strong legs and correspondingly well developed tergocoxal musculature.

Flattening of the meso- and metathorax of cockroaches is achieved by reduction of the epimera (Fig. 7.8). The pleural suture is located obliquely, as are the bundles of the dorsoventral musculature, among which the tergocoxal muscles are the most developed. The pleurocoxal muscles are more abundant than in stoneflies. A small number of powerful tergopleural muscles provide the integrity of the thoracic box, pressing

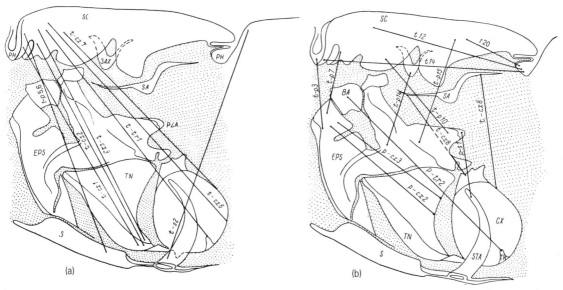

Fig. 7.8 Wing muscles of the cockroach *Periplaneta americana*. Sagittal cross-section of the mesothorax, inner view. (a) and (b) show different muscle groups. The musculature of the metathorax differs from that of mesothorax in that it lacks the muscle *t–p5, 6*, which is a primary wing-levator in the majority of winged insects. This muscle has disappeared in the metathorax due to the very strong development of the tergocoxal muscles.

the tergal margins against the thoracic walls. The tergum is a flat, poorly differentiated plate (Fig. 7.9a). One of the main features of the cockroach tergal structure is the equal development of the scutoscutellar and of the recurrent scutoscutellar sutures. The intersection point of these sutures is marked by the anterior end of the muscle *t20*; in *Leucophaea* its fibres run outside the branch of the recurrent scutoscutellar suture (Fig. 7.9b), whereas in *Periplaneta* they pass inside it (Fig. 7.9a). Nevertheless the anterior end of the *t20* muscle is always attached outside the pseudoscutellum. The cockroach tergal structure is also characterized by a very slender muscle *t14*. The powerful ribbon-like dorsal oblique muscle is attached to the ridge of the scutoscutellar suture over its entire length. The position of the anterior end of *t12* on the scutum and its degree of development bring cockroaches and stoneflies close together. The cockroach pseudoscutellum projects far forward, acting as a stiffening rib which prevents the notum from bending during wing movements. An analogous rib is even more developed in mantids, in which the pterothoracic segments lack any dorsal longitudinal musculature. In cockroaches the dorsoventral muscles are all attached to the anterolateral corner of the notum by a single bundle, whereas in mantids their anterior ends are distributed evenly over the surface of the notum. A ridge along the notum provides its characteristic movement during wing strokes: as the muscle-levators of the wing relax, the notum, as a single rigid plate, moves steadily upwards due to the elasticity of the wing articulations. In cockroaches and mantids wing steering occurs via the anterior notal wing process.

The most characteristic features of the cockroach pterothoracic structure are undoubtedly the poor development of the dorsal longitudinal muscle and the strong development of the oblique one. Because of this some investigators, for instance Tiegs (1955) and Rasnitsyn (1969), have suggested that the dorsal oblique muscles act as substitutes for the functionally weak longitudinal muscles and depress the wings. However, recordings of the electrical activity of the cockroach *Periplaneta americana* wing muscles during flight (Antonova and Brodsky 1977) have

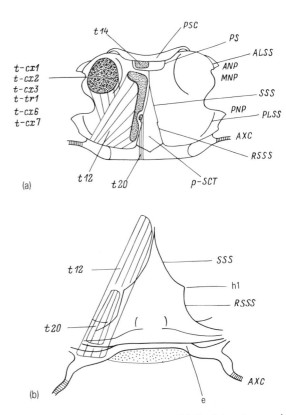

Fig. 7.9 Cockroach metaterga. (a) *Periplaneta americana*. (b) *Leucophaea moderae*, enlarged posterior part of the notum.

shown that the dorsal oblique muscles contract in phase with the tergocoxal muscles and in antiphase with the longitudinal ones, demonstrating that they function as wing levators.

Both notal wing processes connected with the first axillary sclerite are well developed in the axillary apparatus of cockroaches and mantids although the first notal wing process is more strongly developed. The edge of the first axillary sclerite becomes extended from angle *b* to angle *d* (see Fig. 1.10 p.16), and shifts forward to partly enclose the arm, causing the basiradiale to become reduced (Fig. 7.10).

Further modifications in the structure of the cockroach axillary apparatus are associated with improvements in folding the wings flat. The

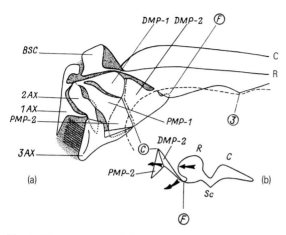

Fig. 7.10 Fragment of the axillary apparatus of the forewing of the cockroach *Periplaneta americana*, showing the arrangement of sclerites and their direction of movement during wing folding. (a) View from above. The border of the clavus is indicated by a broken line. (b) Side view. F is an additional joint which enables the wing to fold flat.

proximal medial plate is divided into two parts at the base of the forewing, so that the posterior part (*PMP*–2) is included in the overturning triangle (Fig. 7.10). Similarly, the distal medial plate is made up of two parts; the apex of the distal part (*DMP*–2) forms an accessory articulation F with the swollen base of *R* such that when the wings are folded and the apex of the distal medial plate is turned downwards the base of *R* is bent upward, allowing the wing to be folded flat. The secondary nature of folding flat is clear here. In the hindwings of cockroaches the distal medial plate forms a characteristically elongated triangle; its overturning at wing folding is so complete that the folded wing becomes horizontal.

When the wings of a cockroach are folded, the leathery forewings completely overlap the membranous hindwings. The hindwings are expanded in the anal area mainly because of the comb-shaped branching of A_2 (see Fig. 1.18, p.28). A_3 is completely eliminated. When in flight, the vannus forms a canopy during the downstroke (Fig. 7.7, 5–9).

Although it is commonly believed that flight does not play an important role in the life of cockroaches, there were no wingless or short-winged forms up to the Tertiary period. However, nowadays cockroaches are obviously heading towards flight reduction. A retiring lifestyle hardly needs to be associated with the ability to fly, although there are still a great many fast-flying forms among species living in the wild.

7.2.2 Changes in the wing apparatus due to adaptations for jumping

The formation of the superorder Orthopteroidea was associated, first of all, with a transition to life in open spaces and to predation; adaptations for jumping have played a decisive role in the success of this group as jumping provided a certain amount of safety from enemies not only for winged individuals but also for nymphs. The history of orthopterans begins with the order Protorthoptera, a small group of rather large insects. Protorthoptera possessed a dorsoventrally flattened body, a short precostal field in the forewings which was uncrossed by veins, and a small anal lobe in the hindwings. These primitive orthopterans presumably lived on plants, spending most of their time in open places. Other orthopterans, and also the Titanoptera and the Phasmodea, probably descended from the Protorthoptera, from which they inherited the characteristic features of hind-motor structure: a flattened body and folding the wings flat.

In modern orthopterans the power for flight is provided by the hindwings. The forewings serve different functions, such as control of flight, stridulation, protection of the folded hindwings, camouflage, and so on. Flight control is the most important function, and has provided orthopterans with 'second wind', their flight becoming faster and more manoeuvrable. In the meso- and metathorax of the locust *Schistocerca gregaria*, the dorsal longitudinal muscles contract before the wings reach the top of the stroke (Fig. 7.11). In the mesothorax they contract long before those in the metathorax, which may explain why the forewings are raised at a shallower angle

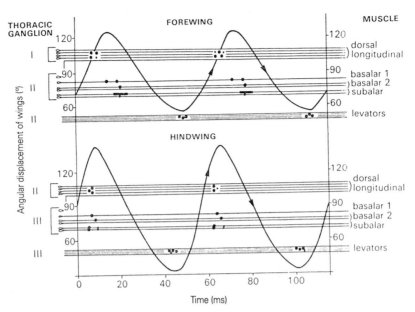

Fig. 7.11 Timing of the firing fore- and hindwing flight muscle motor neurones relative to the wingbeat cycle. Each neurone is shown by a horizontal line, with its origin in the appropriate ganglion on the left-hand side. Each dot on a line represents a nerve impulse occurring at that time; a small dot indicates that an impulse may or may not occur, a large dot that it always occurs. The heavy bar on the forewing subalar muscle motor neurone indicates that firing occurs within this period, but not at a precisely fixed time as with the other units. The curves and the numbers along the ordinate indicate the angular displacement of the wings, where 90° indicates that the wing is horizontal, above 90° that the wings are up, and below 90° that the wings are down (after Wilson and Weis-Fogh 1962).

than the hindwings (Fig. 7.3c). Locust hindwings create lift and thrust, which is why all the muscle-depressors in the metathorax have to contract simultaneously (Fig. 7.11). The situation is quite different in the mesothorax, in which the basalar and subalar direct flight muscles contract in a strictly controlled rhythm, guiding the movements of the forewings. The forewings, in turn, achieve flight control by varying the wing speed and by altering the angle of attack during the downstroke.

The most significant modification of the forewings concerns development or loss of the stridulation apparatus. The veins *CuP*, *PCu*, and the anterior branches of *A* are curved in a characteristic manner when the stridulation apparatus is present. It is believed that the evolution of stridulation was preceded by rustling

of the elevated wings during copulation. The hindwings are broadened by the comb-shaped branching of A_2, and moderate forking in A_1 (Fig. 7.12, see also Fig. 1.17, p.12). A_3 is noticeably reduced, or absent in some species. However, A_3 is well developed in the present-day primitive cricket *Brachytrupes portentosus* (Fig. 7.12).

Adaptations for jumping caused profound changes in the structure and function of the wing apparatus. Jumping hind legs had already appeared in the late Carboniferous Oedischiidae, and the body was flattened laterally. The superposition of lateral flattening on to the usual dorsoventral flattening of most Orthopteroidea caused profound changes in the organization of the pterothorax and also longitudinal bending of the folded forewings. In addition, lateral flattening of the body caused the nymphal

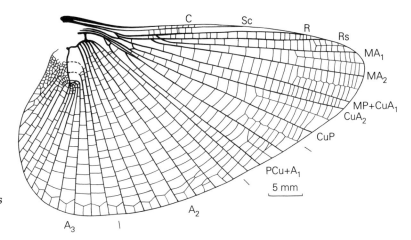

Fig. 7.12 Hindwing of the primitive cricket *Brachytrupes portentosus*, showing a well developed A_3 (after Sharov 1968).

wing pads to turn upwards, on to the dorsal side of the body. The wing pads are orientated such that their costal edges are directed upward and towards the middle line of the dorsum, the hindwing pads overlapping the forewings.

All orthopterans share some features of wing-bearing tergal structure: strong development of the recurrent scutoscutellar suture (Fig. 7.13); poor development of the main hinges of the tergum (*h1*) which do not play an important role in wing motion; unlike cockroaches, the dorsal oblique muscles are weakened or absent; the anterior end of *t20* muscle inserts on the notum outside the recurrent scutoscutellar suture (Fig. 7.14b); the anterior ends of the dorsal longitudinal muscles overlap the notum only slightly, making the parapsidal sutures short; the anterior notal wing process plays the main role in wing operation; and the posterior notal wing process is shifted a long way back because of the broad wing base, and is usually separated from the scutum by a slit (Fig. 7.13a(i)).

The well-developed pseudoscutellum of crickets transfers the power from the contraction of the dorsal longitudinal muscles to the anterior notal wing processes. It can be seen from the sagittal section of the metathorax of the cricket (Fig. 7.13a(ii)) that the pseudoscutellum renders the posterior part of the notum immobile. This organization of the notum means that during the contraction of *t14* the middle

phragma moves backward, arching the scutum upward just in front of the rigid platform of the pseudoscutellum, exactly at the anterior notal wing process. In addition, the leg muscles, which are especially powerful in the metathorax, are attached by a single bundle directly opposite the anterior notal wing process (Fig. 7.13a(i)).

The organization of the terga of katydids and grasshoppers have many features in common (Figs 7.13b and 7.14a). First, development of a membranous area at the site of the transscutal sulcus, which allows the notum to bend along a fracture line when the *t14* muscle contracts. As a consequence of development of the transscutal sulcus, the scutoscutellar suture is greatly reduced, and the dorsal oblique muscles, having lost support, also become reduced. These muscles are retained in the mesothorax of longhorn grasshoppers and katydids but are completely absent in the metathorax, which takes the main load in flight. In the metathorax of grasshoppers and locusts, supplementary hinges (*h4*) have developed at the end of the oblique sutures (Fig. 7.14a) which provide a pivot for the movement of the anterior relative to the posterior part of the notum when *t14* contracts. In these cases, as well as in crickets, the anterior notal wing process carries the main load in wing steering.

The increased role of the anterior notal wing process in wing steering is accompanied by its differentiation into anterior and anteromedian processes (Figs 7.13b(ii) and 7.14a). In the Acrididae, the anteromedian notal wing process is connected to the body of the first axillary sclerite by a fan-like ligament (see Fig. 9.7a).

Lateral flattening of the body and bending of the folded forewing longitudinally along an additional fold led to the alteration of the wing folding arrangement when at rest. Outlining folding appeared, in which the costal edges of the folded forewings are directed downwards (Fig. 7.15a and b).

When the insect is at rest, the forewings cover the hindwings, part of the pterothorax, and the abdomen both from above and on both sides. Each forewing is subdivided into lateral and dorsal lobes. When the wings are folded, the lateral lobe is held close to vertical, while the dorsal lobe is usually horizontal. The left and right wings partly overlap, usually with their dorsal lobes. In the Ensifera two different wing positions may be adopted when at rest: slightly depressed (Fig. 7.15a), and elevated (Fig. 7.15b). When the wings are held elevated, the base of the abdomen is enclosed on both sides by the forewing bases, and the apex of the abdomen is free. The folded hindwings sit in the space between the forewings and the dorsal surface of the body. When the wings are depressed, the whole abdomen is enclosed on both sides. Slightly depressed forewings are characteristic of lower Ensifera, combined with a poorly developed anal lobe of the hindwings.

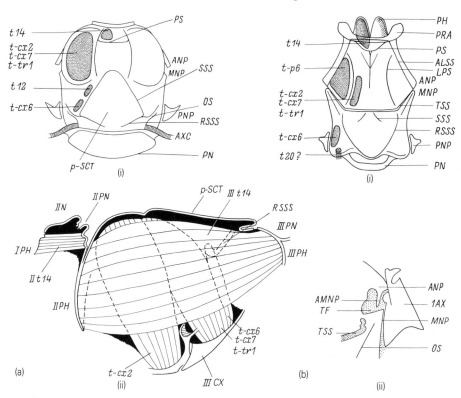

Fig. 7.13 Structure of the pterothorax in Orthoptera Ensifera. (a) The cricket *Gryllus bimaculatus*: (i) general view of the metathorax from above and (ii) sagittal cross-section of the meso- and metathorax showing the huge tergocoxal muscles. (b) The bush cricket *Decticus albifrons*: (i) general view of the metathorax from above and (ii) enlarged region of the anterior notal wing process.

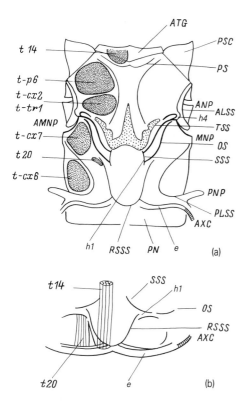

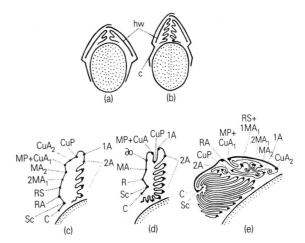

Fig. 7.15 Transverse section through the abdomen of various orthopterans, showing the hindwings folded beneath the elytra when the forewings are (a) depressed and (b) elevated; and position of the folded hindwing in (c) Oedischiidae, (d) Tettigoniidae, and (e) Gryllotalpidae. By kind permission of Gorochov; vein nomenclature after Sharov (1968).

Fig. 7.14 Metatergum structure of the locust *Locusta migratoria* (Orthoptera, Caelifera). (a) General view from above and (b) enlarged posterior part of the notum showing the site of attachment of the *t20* muscle.

Orthoptera in which the vanni of the hindwings are not expanded fold their hindwings so that the base of the vannus is held vertically (Fig. 7.15c). The hindwings of Tettigoniidae are folded similarly (Fig. 7.15d), although their vanni are expanded much more. In crickets and mole crickets the area of the hindwing from C to CuA_2 (Fig. 7.15e) is held close to horizontal when folded, which makes the folded wings appear flat. The secondary nature of such folding is obvious.

In crickets and mole crickets the jugal fold begins, as usual, at the vertical hinge, goes round a small proximal medial plate, and reaches the anal area further down the base of the radiating branches of A_2 (Fig. 7.16). There is

also an anal fold; it is curved distally and, after going round a large distal medial plate, turns down, cutting across the bases of the anal veins. The veins are curiously curved where it runs. The comb-like radiation of A_2 branches from a broad common base is an aerodynamic specialization which makes compact bending of the vannus somewhat difficult. When the hindwing is folded a narrow triangle, bordered exteriorly by a jugal fold, turns over, and the whole wing plate, including the remigium and the vannus, deflects backward and folds along the vannal folds. The anal fold serves as the interior border of the part of the wing which deflects backwards (Fig. 7.16). In the Acrididae the anal fold is in the postcubital area, while in Tettigoniidae it reaches the wing edge in the medial area. The jugal fold, if retained, does not participate in wing folding in these insects.

Flight is relatively subordinate to jumping in orthopterans. Wingless forms are common in many groups. The hindwings become reduced first, whereas the forewings, which may serve

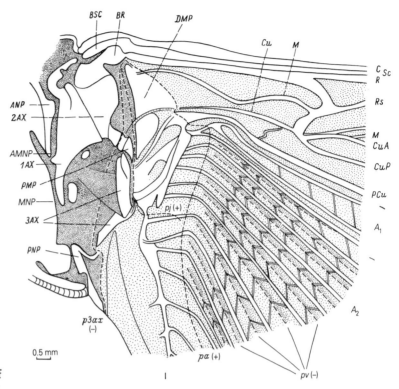

Fig. 7.16 Hindwing base of
the mole cricket *Gryllotalpa
gryllotalpa.*

as protective or signalling organs, only become reduced as a last resort. Many species exist in two forms: fully-winged and short-winged. Among fossilized Ensifera forms with partially reduced or absent wings are rare, and are known only from the Kainozoic era.

Most orthopteran trivial flight comprises relatively high speed movement along a rectilinear path. It is when migrating in giant swarms that locusts reveal their flight abilities best. Observations demonstrate the remarkable precision with which a flying locust is able to correct for yaw by maintaining the relative airflow strictly parallel with its body's longitudinal axis, not only in straight flight but also during turns (Rainey 1985). Thus, for example, one particular cine sequence showed a 70° change of flight orientation, occurring progressively within a period of 1.2 s, passing through the upwind direction

and so resulting in a 180° change in direction of movement relative to the ground. Throughout this sequence the changing orientation of the locust, in relation to the wind, remained very closely consistent with its track relative to the ground.

Giant swarms of locusts cause considerable damage to agriculture. This ability, together with their large size, and the ease with which they can be reared has made two species of locusts, *Schistocerca gregaria* and *Locusta migratoria*, popular subjects for the investigation of insect flight. The 'Anti-locust Research Centre' was founded in the United Kingdom specifically for this purpose. Weis-Fogh and Jensen (1956) carried out investigations at the Centre that stimulated a systematic and thorough study of insect flight.

7.3 Conclusion

The main direction of specialization of the polyneopteran wing apparatus was towards living in crevices and cavities in the litter. Dorsoventral flattening of the body was accompanied by a diminishing role for the dorsal longitudinal muscles as principal wing depressors; this role was taken by the musculature of the basalar complex so that the main load in wing steering fell on the anterior notal wing process. A new tergal structure appeared, the pseudoscutellum, which shifted the strain of deforming the notum on to the anterior wing processes. Wing kinematics underwent considerable modifications; plecopteran quasi-synchronous kinematics was substituted for by hind-motor kinematics in the Blattodea and Orthoptera. A low stroke amplitude of the forewings is usual in hind-motor kinematics. The forewings began to operate in the lower hemisphere, while in contrast, the stroke amplitude of the hindwings was large. The pattern of wing deformation remained the same as in Protoptera—supinational twisting of the forewings and gradual supination of the hindwings (except that unlike in the Protoptera, the folds were shifted backwards relative to the veins). In some stoneflies a more progressive type of hindwing deformation evolved. This supinational bending starts from an abrupt wing tip deflection and continues as a longitudinal bending of the wing blade along the fold of the radial sector.

The transition from the primitive folding of the Protoptera to flat folding was accompanied by elytrization of the forewings and considerable expansion of the hindwings. At wing folding an expanded anal fan deflects along the anal fold; the jugal fold is retained in stoneflies and has nearly disappeared in cockroaches and orthopterans. Compact packing of the anal fan inside the deflected lobe was provided by the newly evolved vannal folds. The development of jumping in Orthoptera triggered a chain of events, such as lateral body flattening and the appearance of outlining wing folding. This process, which in some Ensifera resembles the flat wing folding, conceals a complex mechanism for packing the broad hindwings.

8

From functionally four-winged flight to functionally two-winged flight

. . . this bug with gilded wings,
This painted child of dirt, that stinks and stings.

Alexander Pope

Unlike in the Polyneoptera, the course of evolution of the Para- and Oligoneoptera proceeded, first of all, in the direction of flight improvement, and was accompanied by reconstruction of the wing mechanics. In his book *'Historical development of the Class Insecta'*, Rasnitsyn (1980d) offers the following description of the common ancestor of Para- and Oligoneoptera: 'The common ancestor of the cohort which possessed primitive features of both the Blattinopseida and Hypoperlida presumably resembled modern alderflies (Sialidae), but was probably larger (3–5 cm in length). The mouthparts were of the biting type, non-specialized, the thorax had a strongly invaginated sternum, the wings were of elongated oval shape, the forewings might have been slightly thickened, the hindwings probably had an expanded anal area'. This ancestor could have been a member of the superorder Caloneuroidea. Its probable descendants, aside from the holometabolous insects (Oligoneoptera), were the superorder Hypoperloidea which comprised the single order Hypoperlida. Outwardly, the Hypoperlida resembled the Neuroptera and Mecoptera—they had the same light, slender body, membranous wings of a similar shape but with characteristically different venation. When at rest the wings were held either flat or roof-like. Thus the presence of two pairs of nearly similar,

uncoupled wings was usual in all insects close to the common ancestors of the Para- and Oligoneoptera. The wing folding arrangement was not yet settled, and presumably included features of both flat and roof-like arrangements. Depending on the conditions of fossilization, some species of Hypoperlida show roof-like wing folding, while others fold the wings flat. True roof-like folding occurred in the Blattinopseida, whose forewings possessed an expanded costal field (Fig. 8.1). However, elongated narrow wings with a weakly expanded costal field, which were characteristic of the Caloneurida could hardly be held in roof-like fashion. In a lower Permian representative of this order, *Paleuthygramma tenuicorne* (Fig. 8.2), the wings were folded flat over the abdomen. In any case, this evidence shows that the folded wings were not locked on the back, so that their position was easily altered by fossilization conditions.

The principal feature of the wing apparatus of the insects discussed above is the presence of two pairs of equally developed and independently functioning wings. Among contemporary insects, functionally four-winged flight is usual in the Neuroptera, Mecoptera, and Megaloptera, and also in primitive caddisflies and moths with uncoupled wing pairs.

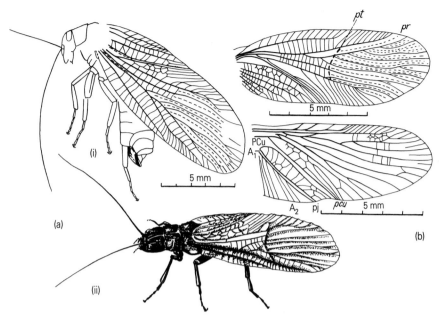

Fig. 8.1 Representatives of the order Blattinopseida. (a) *Glaphirophlebia uralensis* general view: (i) a fossil and (ii) reconstructed. (b) Fore- and hindwings of *G. subcostalis*. The resemblance to alderflies is emphasized by roof-like wing folding (after Rasnitsyn 1980*a*).

8.1 The flight of functionally four-winged insects

In all functionally four-winged insects the wing pairs go down synchronously and move up asynchronously, the hindwings moving with the greater amplitude. Wingbeat frequency is relatively low. The stroke planes of both wing pairs are nearly perpendicular to the longitudinal body axis. The body angle is usually large relative to the horizontal. The forewings of the antlion *Myrmeleon formicarius* rise at a steeper angle than the hindwings, and then clap their dorsal surfaces together (Fig. 8.3, 1 and 2). When the hindwings reach the end of the upstroke, both pairs pronate synchronously (Fig. 8.3, 3 and 4) and the downstroke begins. During the downstroke the hindwings slightly anticipate the forewings. As they approach the bottom of the stroke, the forewings decelerate, whereas the hindwings continue moving further down to a larger angle (Fig. 8.3, 11). At the end of the upstroke the hindwings again come up

with the forewings. The greatest phase shift, one quarter of a stroke cycle, is observed during elevation of the wings (Fig. 8.3, 14 and 15). In some functionally four-winged insects such as alderflies and scorpionflies, the wingstroke amplitude of the forewings is even less (Fig. 2.7d and e, p.39) but they always oscillate in the upper hemisphere, this feature strikingly distinguishing the wing kinematics of these insects from those of hind-motor insects. In hind-motor insects the forewings oscillate either in the lower (cockroaches, mantids) or in the middle (orthopterans) hemisphere, but they never reach the level of the hindwings at the end of the upstroke.

Wing deformation of functionally four-winged insects was studied by Grodnitsky and Kozlov (1985) in moths of the families Micropterigidae and Eriocraniidae, by Ellington (1984) in the green lacewing *Chrysopa carnea*, and also by

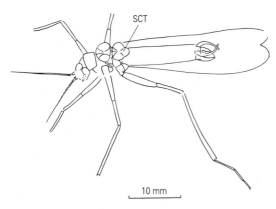

Fig. 8.2 Representative of the order Caloneurida *Paleuthygramma tenuicorne*. The resemblance to a mosquito is emphasized by flat wing folding (after Rasnitsyn 1980*a*).

Ivanov (1985) in rhyacophilid caddisflies. Finally, Ennos and Wootton (1989) studied deformation of the wings during the stroke cycle of the scorpionfly *Panorpa germanica*.

The positions of the wings of the alderfly *Sialis morio* in the course of the stroke cycle are shown in Fig. 8.4. The forewings begin the downstroke pronated (Fig. 8.4, 1), then at the middle of the downstroke the wings supinate slightly (Fig. 8.4, 2). At this point pronounced grooves produced by the convex veins, the concave cubital and postcubital folds, and the radial sector fold appear on the wing surface. The membranous areas between the veins are concave (Fig. 8.4, 2). This profile is maintained until the wing decelerates at the end of the downstroke.

As the forewing decelerates (Fig. 8.4, 4), its tip points downwards and the wing plate bends ventrally, forming a sharp diagonal flexion line.

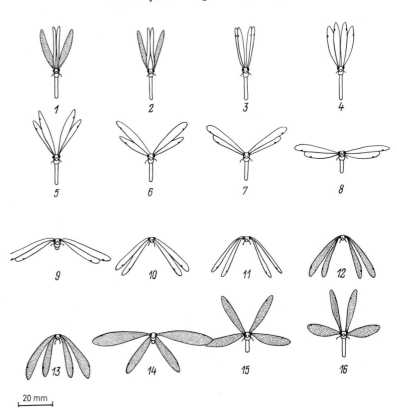

Fig. 8.3 Flight of the antlion *Myrmeleon formicarius* (front view). Consecutive film tracings (1–16) from a single wingbeat. Stroke parameters: $n = 15$ Hz, $A_{fw} = 130°$, $A_{hw} = 150°$, $b_{fw} = 80°$, $b_{hw} = 70°$, $B = 0°$. The lower surfaces of the wings are shaded.

Fig. 8.4 Deformation of the wings during the stroke cycle of the alderfly *Sailis morio*, as seen from a series of photographs taken obliquely from above. Stroke parameters: $n = 30°$, $A_{fw} = 100°$, $A_{hw} = 135°$, $b_{fw} = 70°$, $b_{hw} = 70°$, $B = 20°$.

On the wing plate, this line projects postero-basally from the apex of the subcostal vein, then, after meeting the medial vein, turns backwards and outside, and runs along the anterior cubitus up to the rear edge of the wing, as is shown in Fig. 2.13 (p.45). There are numerous grooves on the deflected part of the wing. The posterior edge of the clavus is also inclined downwards. From this moment, the wing starts turning its leading edge upwards. First, the wing bends longitudinally over the radial sector fold, taking with it the deflected part of the wing. Second, once the longitudinal bending occurs, the wing plate straightens, and its tip returns to its usual position (Fig. 8.4, 5). At the same time the wing bends along the cubital fold, the clavus flattens, and the remigium turns with its anterior edge upwards.

At the upper point of the stroke, the hindwings are more steeply raised than the forewings. Pronation of the hindwings has some

peculiarities and proceeds more slowly than that of the forewings. During pronation the wings first bend along a line which coincides with the medial fold; with time, this line gradually shifts backwards, towards the wing base (Fig. 8.4, 1). At the middle of the downstroke the posterior part of the hindwing is stretched out and is depressed slightly, thus producing a canopy (Fig. 8.4, 2) which is retained until the end of the downstroke (Fig. 8.4, 3).

When the hindwing stops at the lowest point of the trajectory, its tip deflects downwards as in the forewing. On the wing plate, the flexion line projects backward from the apex of *Sc*, and after meeting the anterior cubitus, turns along it to the rear edge of the wing (see Fig 2.13, p.45). At the moment when the wing tip deflects, the entire posterior part of the wing is folded up like a fan (Fig. 8.4, 4). Then supination of the wing plate starts, beginning from the fold of the radial sector, causing the wing tip to straighten. At the same time, a deep groove forms along the cubital fold, and the wing sector which includes all branches of *CuA* and *CuP* becomes convex (Fig. 8.4, 5). In the anal region of the wing a zigzag profile is formed by convex ridges and concave grooves which run at an angle to the trailing edge. By the end of the upstroke the hindwings leave the forewings behind (Fig. 8.4, 6), and finally clap their dorsal surfaces together.

As is clear from the description above, deformation of alderfly wings is of the supinational bending type: supination begins with tip deflection downwards along a sharp diagonal flexion line and continues as a longitudinal bending of the wing blade along the fold of the radial sector. However, at the start of supination the forewing twists at its base (Fig. 8.4, 4), which is reminiscent of supinational twisting. The latter starts with a sloping wing tip deflection downwards; in this the wing plate bends over the line of anastomosis which crosses the remigium obliquely, from the apex of *Sc* to the point where the cubital fold reaches the rear edge of the wing. Both regions are most flexible at the margin, the bending point of the leading edge being located distally relative to the corresponding point of the rear margin (see Fig. 1.15b). With the development of this deformation, the point at which convex bending of the leading edge occurs is gradually shifted apically, whereas that of the rear edge is shifted basally. It has the appearance of a wave passing along the wing surface, resulting in complete supination of the costal edge. It is just when this movement occurs that deformation of the alderfly forewing begins. Because of this, one needs to consider that in the forewing of the alderfly the anterior part of the flexion line, which is located obliquely on the wing membrane (see Fig. 2.13, p.45), corresponds precisely to the line of anastomosis. It is therefore obvious that wing deformation in functionally four-winged insects has developed on the basis of supinational twisting; the posterior end of the line of anastomosis has shifted tipward, and the line along which the tip deflects has become characteristically pointed (Fig. 2.13). Where the wing shape is unchanged, the line

of tip deflection has remained oblique, and the wing therefore experiences the supinational twisting during the upstroke. For instance, in caddisflies of the family Rhyacophilidae, both pairs of wings undergo strong torsion during the upstroke. Supinational twisting is also retained in some form in the forewings of scorpionflies, alderflies, and many species of Neuroptera. Nevertheless, the main type of wing deformation in functionally four-winged insects is supinational bending.

Although the wing pairs do not move synchronously and have different stroke amplitudes, the whole vortex formation pattern in the flight of functionally four-winged insects is close to that of stoneflies. Flying *Hemerobius simulans* and moths of the families Micropterigidae and Eriocraniidae leave a chain of coupled vortex rings in the air. Ivanov (1990) studied the pattern of vortex formation during the flight of the caddisfly *Rhyacophila nubila*, and showed that during the downstroke the wings of one side act as a common aerodynamic surface, even though there may be a gap between the fore- and hindwing, which are moving in one plane (Fig. 8.5a). During the upstroke, airflow rushes through the gap between the fore- and hindwing (Fig. 8.5b). At that moment the ventral vortex is located near the tip of the hindwing. As they elevate, the hindwings catch up with the forewings so that the gap between them is diminished, but nevertheless the air continues to flow through the space between the wings. This appears to compensate for its deleterious effects during wing elevation in reducing mean lift. The air which passes between the fore- and hindwing at the beginning of the upstroke creates additional positive lift, which acts on the hindwings as they continue to move downwards (Fig. 8.5b and c). The downward deflection of the forewing tip changes the orientation of the tip vortex, turning it horizontally and causing the ventral vortex to appear (Fig. 8.5a).

Experimental study of the vortex formation pattern during the flight of functionally four-winged insects has shown that the difference in stroke amplitude between the wing pairs does not affect the resultant vortex formation, since the two wing pairs pronate and supinate

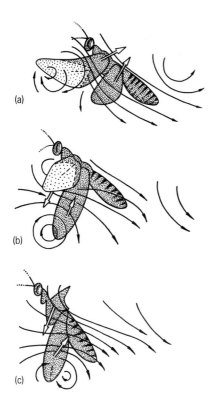

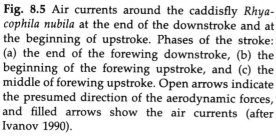

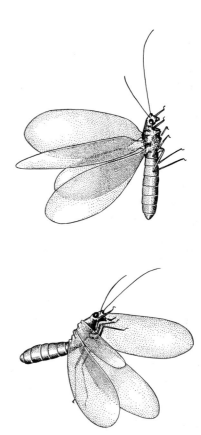

Fig. 8.5 Air currents around the caddisfly *Rhya-cophila nubila* at the end of the downstroke and at the beginning of upstroke. Phases of the stroke: (a) the end of the forewing downstroke, (b) the beginning of the forewing upstroke, and (c) the middle of forewing upstroke. Open arrows indicate the presumed direction of the aerodynamic forces, and filled arrows show the air currents (after Ivanov 1990).

Fig. 8.6 Posture of the green lacewing as it flies upwards (after Dalton 1975).

synchronously, and then the vortices are shed from the wings. So, two pairs of wings moving differently, create a common system of coupled vortex rings. Due to the differences between the up- and downwards movements just discussed (Fig. 8.5b and c), the force impulse created during the downstroke has to be greater than that generated during the upstroke. Thus the insect generates a smaller diameter ring during the downstroke than during the upstroke, resulting in the formation of a vortex wake whose

sagittal section forms a slanted vortex street.

As in stoneflies, the hindwings interact more actively with the airflow than the forewings, and this might be one of the reasons why hindwing stroke amplitude is larger than that of the forewings. Another reason might also be that the forewings, when at the middle of the downstroke, meet the oncoming airflow first and decelerate more quickly than the hindwings. When the forewings stop, the middle phragma which has been shifted forward slightly returns to its original state, thus causing an additional depression of the hindwings. Unlike in stoneflies, in functionally four-winged insects the contraction of dorsal longitudinal muscles in the mesothorax shifts both the anterior and the middle phragmas.

Functional four-wingedness gives rise to 'yawing' flight—slow flight with poor stability and bad manoeuvrability. The flights of *Micropterix calthella* moths from flower to flower do not take the shortest path; because of the yawing movements their trajectory is complicated and includes many loops. Rapid alterations in flight direction are not accompanied by turning the longitudinal body axis in the direction of motion. Climbing with the head facing upward is very characteristic of green lacewings (Fig. 8.6). This habit, as well as other manoeuvres such as somersaults and spiral flights performed at low speed, are highly typical of other functionally four-winged insects hunting among vegetation.

8.2 The origin of functionally two-winged flight

Functionally four-winged flight was the first step in the optimization of the interaction between the wing pairs in the Paraneoptera-Oligoneoptera evolutionary branch. Considerable difficulties arise in the coordination of muscle contraction in the meso- and metathorax if the wing pairs are depressed synchronously and elevated asynchronously. The dorsal longitudinal muscles, as has already been mentioned, have to contract concurrently, but because of the asymmetry of meso- and metathorax they act differently in both segments causing the dorsoventral muscles to contract independently, and allow asynchronous elevation of the wings. The transfer of steering of both wing pairs to the mesothorax considerably simplified the whole organization of the flight system, and provided a starting point for the development of physiological mechanisms which allow high frequency wing oscillations. The forewings became dominant, and functional two-wingedness appeared. The consequent decrease in hindwing stroke amplitude was compensated for by their becoming broader in caddisflies and moths.

Functionally two-winged flight has also passed through several stages of development: first, synchronous movements of the wing pairs were provided by a common rhythm of muscle contractions, and then various coupling mechanisms evolved. This enabled synchronization to be more complete and continuous throughout the ascending part of the trajectory. But to begin with it was only during the downstroke that the coupling mechanisms permitted synchronous wing movements. For instance, in some caddisflies the wings uncouple and elevate independently during the upstroke.

In the Homoptera Sternorrhyncha and Psocoptera the coupling mechanism enables the hindwing to rotate relative to the trailing edge of the forewing, thus abruptly altering its profile at various stages of the stroke. The hindwing rotation is greatest at the stroke extremes. The best synchronicity is provided by rigid coupling of the fore- and hindwing, so that they present a constant outline of common aerodynamic plane. The swiftest hymenopteran flyers have strongly coupled wings. Their wings are coupled together by hooks and loops whereby the hindwing can be tilted against the forewing by muscle power, thus working as a forewing trailing-edge flap; changing the flap angle alters the lift and drag, and high-speed films of bees show that this mechanism is used to turn the body to the optimal landing position. With the appearance of coupling mechanisms, neural coordination of the fore- and hindwing action was substituted for mechanical coordination, which improved the reliability of the functioning of the wing apparatus and allowed increased wingbeat frequency.

8.3 Evolution of the paraneopteran wing apparatus

The three orders at the base of the Paraneoptera, the Hypoperlida, Blattinopseida, and Caloneurida, constituted typical functionally four-winged insects. From one of these orders, the Hypoperlida, a branch diverged and gave rise to the Psocoptera. An extinct suborder Permopsocina

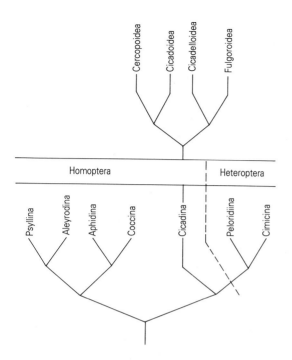

Fig. 8.7 Basic phylogenetic scheme of the Rhynchota (after Emelyanov 1987).

included the most archaic psocids which had nearly identical wings, which were uncoupled during flight. However, a fore- and hindwing coupling mechanism was already present in psocids, and functional two-wingedness arose. The psocid coupling device consisted of hooks on the ventral surface of the forewing and a lip-like fold along the anterior margin of the proximal part of the hindwing which was turned in a dorsal direction. The catch on the forewing is rather primitively constructed, whereas its orientation, position, and mode of functioning are quite close to those of the Homoptera and Heteroptera.

The most characteristic feature in the evolution of the Paraneoptera is their transition from feeding on the generative organs of plants to feeding on vegetative tissues (and from there sometimes to predation) or to mycophagy, as in recent psocids. These feeding specializations led to the decreased importance of flight for these

insects, which, in turn, caused profound changes in the insects' appearance. The psocid body was already more compact and bulky, and it has remained this in most of the Paraneoptera. The transition to a low-mobility lifestyle caused wing reduction in some cases; where they are retained, they can be locked in a roof-like position on the back.

The Cohort Paraneoptera, apart from the extinct orders discussed above, includes the Psocoptera, Thysanoptera, and Homoptera plus Heteroptera (Hemiptera). The Homoptera and Heteroptera are separated off into the Rhynchota because of the transformation of their mouthparts into a segmented sucking proboscis. The first branching into Rhynchota is believed to have taken place between the Sternorrhyncha and the common stem of the cicadiform Homoptera and Heteroptera (Fig. 8.7). One branch, the Sternorrhyncha, comprises minute and very minute insects. The flight of minute insects is based on quite different aerodynamic principles and will be discussed in Chapter 10. This is also true for thrips.

The Homoptera Auchenorrhyncha and Heteroptera are distinguished from the Sternorrhyncha by a number of features of the organization of the wing apparatus. First, the hindwings are equipped with a broad, similarly structured vannus; second, the basal lobe of the clavus is drawn into a peak; third, the posterior border of the pronotum forms a peak that covers a considerable part of the mesonotum and the bases of the forewings; fourth, the claval edges of the folded forewings go into the oblique grooves that border the interalar triangle of the mesonotum, the so-called 'scutellum'; fifth, the fore-and hindwings are coupled by interacting lip-like outgrowths of the wing during flight, whereas in the Sternorrhyncha, in the hindwing, coupling is provided by hook-like formations of chetoid origin.

8.3.1 The Psocoptera

A powerful dorsal oblique muscle and a correspondingly well-pronounced scutoscutellar suture of the tergum is usual on the wing-bearing segments of booklice and psocids, as well as for most of the Paraneoptera (Fig. 8.8a). A tendency towards the fusion of the anterior and median notal wing processes, which results in the lack of a transscutal sulcus is also characteristic. A platform bordered by the oblique suture and the edge of scutellum is well developed on both sides of the scutoscutellar suture; this platform is used to lay the folded wing on the back. Both the scutoscutellar and the recurrent scutoscutellar sutures are clearly pronounced and the main hinges of the tergum (*h1*) developed at their point of intersection. The scutellum has a characteristic shape but lacks the muscle *t13*; it is not clear, however, whether this is associated with minute size of the insects or whether *t13* is primarily lacking. Almost the entire surface of

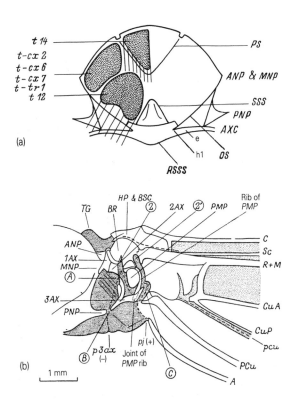

(a)

(b)   1 mm

Fig. 8.8 Details of the wing apparatus structure of the booklouse *Psococerastis gibbosus*: (a) dorsal view of mesotergum, and (b) the forewing base. 2' is a novel additional vertical hinge in the wing base of the Para- and Oligoneoptera.

the scutum is occupied by the sites of attachment of three groups of muscles: *t14*, *t12*, and the tergocoxal muscles (Fig. 8.8a). Muscle *t-p5, 6* is absent.

In the course of evolution the ancient roof-like arrangement of folded wings gave rise to the flat folding and also to true roof-like folding. True roof-like folding had presumably appeared already in the common ancestors of the Para- and Oligoneoptera, but it was not sufficiently well perfected, and did not provide reliable locking of the folded wings. Refinement of the roof-like arrangement also occurred in the Para- and Oligoneoptera, but was achieved through different mechanisms. Changes in wing folding mechanisms are already apparent in booklice: the proximal medial plate becomes immobilized, and a rib which hinges with the distal arm of the third axillary sclerite is formed along its external edge (Fig. 8.8b). As the muscle *t-p13* contracts, the distal arm of the third axillary sclerite, as is usual, moves upwards and moves closer to the body. As the wing moves backwards the anterior end of the rib of the proximal medial plate rests against the basiradiale, making the wing plate bend longitudinally along the distal branch of humeral fold. In booklice the posterior arm of the third axillary sclerite rests upon the subalar sclerite, which is believed to take part in wing folding (Badonnel 1934). Contraction of the *t-p13* and subalar muscles results in partial overturning of the distal arm of the third axillary sclerite, which noticeably shifts backwards because the joint B ceases to act as a hinge, and therefore is depressed with the subalar sclerite.

Little is known about the flight of booklice and psocids. Two pairs of membranous wings are folded in a roof-like manner over the abdomen when at rest, and sometimes the wings are reduced. The diverse wing structure of booklice may indicate a moderate variety of flight ability.

8.3.2 The Homoptera Auchenorrhyncha

The main features of the mesothorax organization of the cicadiform Homoptera are the following: hypertrophy of the dorsal oblique muscle, fusion of the anterior and median notal

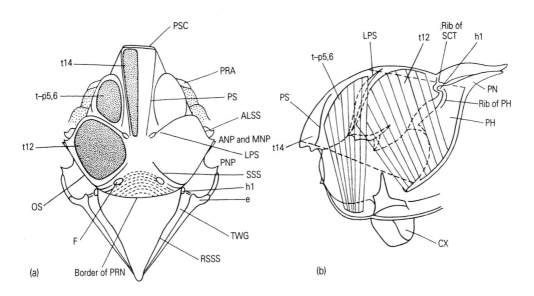

Fig. 8.9 Mesothorax of *Aetalion reticulatum* (superfamily Cicadelloidea): (a) dorsal view and (b) sagittal cross-section showing the huge *t12* muscle; the *t14* muscle is shown by a broken outline.

processes, expansion of the posterior part of the scutellum and its transformation into a peak, and formation of a tergal wing groove into which the posterior edge of the folded forewing fits.

In *Aetalion reticulatum* which can be regarded as the most primitive representative of the superfamily Cicadelloidea (Fig. 8.9), the powerful muscle *t12* is shifted forward into the area of the anterior and median wing processes, and the median part of the scutoscutellar suture is therefore reduced (Fig. 8.9a). The area of attachment of this muscle to the notum is bordered by the lateral parapsidal suture from anteriad and by the branch of the scutoscutellar suture from posteriad. The posterior notal wing process is retained. The recurrent scutoscutellar suture is developed to a greater extent than the scutoscutellar one. The main hinges of the tergum are located at the points where they intersect, and oblique sutures radiate out from them. The mesonotum is covered by the pronotum; its posterior border, in the midpart of the notum, is marked by a sulcus which looks like the continuation of an oblique suture and passes between the main hinges (Fig. 8.9a). When

Aetalion's wings are folded, part of the scutellum is visible from above, surrounded by the posterior border of the pronotum and by the recurrent scutoscutellar suture, which is drawn out backwards to a point. This region of the scutellum is named the interalar triangle. The brim-like edge of the scutellum is quite long; the posterior edge of the folded forewing fits into a groove between it and a wall of the interalar triangle. The area of the scutellum under the pronotum is covered in a network of small creases. There are small oval fossae posteriad from the branches of the scutoscutellar suture, exactly opposite the sites of attachment of the dorsal oblique muscles (Fig. 8.9a). An endoskeletal ridge projects away from each fossa, passes beneath the notum surface and articulates with the phragma rib by the main hinge (Fig. 8.9b). The reason for this hinge articulation becomes clear if we consider that the posterior ends of the dorsal oblique and longitudinal muscles are inserted on the phragma (Fig. 8.9b). As the *t12* muscle contracts, the inferior angle of the phragma moves forwards and upwards, while the rear edge of the postnotum, which points backwards in a beak-shape, moves downwards and backwards. Contraction of the dorsal oblique muscle therefore causes the removal of the brim-like edge of the scutellum from the outer edge of the interalar triangle, expansion of the tergal wing groove, and thence release of the forewing from its resting position.

In *Tibicen plebejus* the dorsal oblique muscle is even more strongly developed, and consists of four bundles (Fig. 8.10a (i)). Since the region of its attachment to the notum is shifted forward, the branches of the scutoscutellar suture have been subject to further reduction; only small fragments are retained on the upper surface of

the notum (Fig. 8.10a(i)). An interalar triangle does not occur as the pronotum border ends on the prescutum. The tergal wing grooves are well developed, and are shorter and broader than in *Aetalion reticulatum*. The rib of the scutellum that corresponds to the scutoscutellar suture is well pronounced and can be easily seen both from the outside (Fig. 8.10a(ii)) and the inside (Fig. 8.10a(iii)). Fossae homologous to those in *Aetalion reticulatum* are present on the posterior half of the notum.

In the most primitive representative of the superfamily Cicadoidea, *Tettigarcta tomentosa* (Fig. 8.10b), the lateral parapsidal sutures are

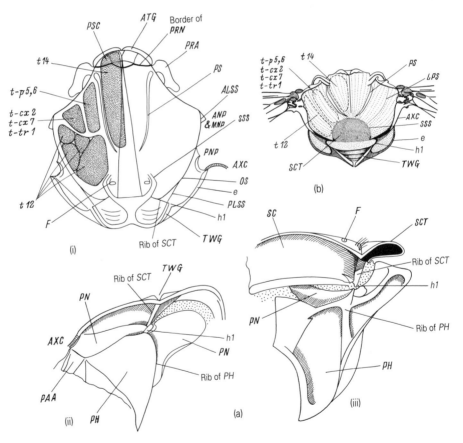

Fig. 8.10 Mesothorax of cicadas of the superfamily Cicadoidea. (a) *Tibicen plebejus*: (i) dorsal view, posterior part of notum as seen from (ii) outside and (iii) inside, the head to the left. (b) *Tettigarcta tomentosa*, dorsal view (after Evans 1941), with the sites of muscle attachment outlined.

present and the circo-triangular scutellum is well developed. The recurrent scutoscutellar suture is swollen, and forms a peak that hangs over the tergal wing grooves. There is an interalar triangle whose topography exactly parallels that of *Aetalion reticulatum*. Evans (1941) did not show fossae in *Tettigarcta* although many authors do so when describing the morphology of Homoptera for taxonomic purposes. Fossae homologous to those of *Tibicen plebejus* and *Aetalion reticulatum* are present in many Homoptera Auchenorrhyncha. Sometimes they may be weakly marked and only distinguishable from the background by their colour. In the green leafhopper, *Cicadella viridis* the fossae are located anteriad from the suture which Savinov (1983) designated the scutoscutellar suture. There are no fossae in the froghopper *Aphrophora salicina* (Savinov 1986). However, they are present in many other members of the superfamily Cercopoidea; and as in the green leafhopper, they are located anteriad from the suture. The fossae in question are clearly traces of the previous attachment of *t13*, which is indicated by their position relative to the sites of attachment of the *t12* muscle and the branches of the scutoscutellar suture in primitive cicadiform Homoptera. The suture behind these fossae corresponds to the anterior border of the peak in some leafhoppers, formed by the expanded recurrent scutoscutellar suture. The suture in question was already well pronounced in *Tibicen* and *Tettigarcta*, and also in *Aetalion*, where it coincides with the posterior border of the pronotum. The true scutoscutellar suture is situated anteriad from the fossae, but in some Homoptera Auchenorrhyncha it is completely or partly reduced. If the fossae are really the former sites of attachment of the scutoscutellar muscle, cicadiform Homoptera must have possessed a true scutellum—the same one as in the Oligoneoptera, whose characteristic feature is the presence of the *t13* muscle. The true scutellum was lost by the ancestors of cicadiform Homoptera, presumably because the upper ends of the *t12* muscle shifted forward and also because of the formation of the apparatus for locking and releasing the folded wings. That the ancestors of the Rhynchota and Psocoptera had a true scutellum is indirectly indicated

by the presence of a characteristically-shaped shield, with corresponding orientation of the branches of the scutoscutellar suture (so as to be convenient for attachment of the anterior ends of the *t13* muscle) in psocids (Fig. 8.8), aphids, psyllids, and other Sternorrhyncha (Taylor 1918; Weber 1929; Matsuda 1970). Illustrations of the terga of extinct Caloneurida (Fig. 8.2, p.134) also suggest that these insects already possessed a true scutellum.

The structure of the Homoptera Auchenorrhyncha wing base exemplifies the developmental tendencies of booklice and psocids. The anterior and the median notal wing processes have come close to each other and form a single process which articulates with the first axillary sclerite. This has led to reduction of the arm of the first axillary sclerite, while the length of its body from angle *b* to angle *d* (see Fig. 1.10, p.16) has been stretched (Fig. 8.11). The proximal medial plate has become completely immobile and fused with the second axillary sclerite, the distal medial plate is divided into *DMP-1* and *DMP-2*, and the joint B has completely lost its role in wing folding (Fig. 8.11).

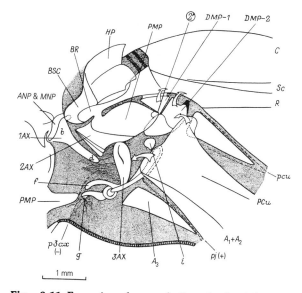

Fig. 8.11 Forewing base of the cicada *Tibicen plebejus*. Arrows indicate the direction of sclerite movement during wing folding.

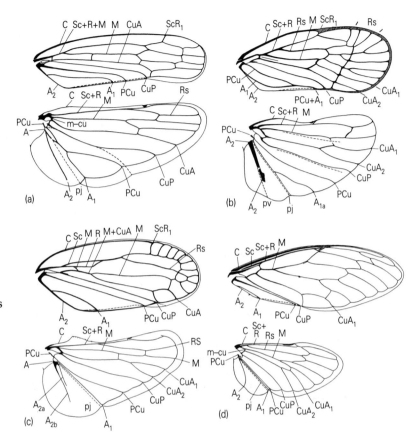

Fig. 8.12 Fore- and hindwings of different cicadiform Homoptera Auchenorrhyncha (after Emelyanov 1987, with lettering partly changed).
(a) Superfamily Cicadelloidea.
(b) Superfamily Fulgoroidea.
(c) Superfamily Cercopoidea.
(d) Superfamily Cicadoidea.

Specialization for locking roof-like folded wings on the back has affected the structure of the clavus of the forewings; as a result of the basal fusion of A_1 and A_2 (Fig. 8.12), the posterior edge of the clavus has acquired the shape of the tergal wing groove. Dworakowska (1988) suggested that in some cicadiform Homoptera, A_3 is retained at the base of the jugum. Presumably this was the background against which the expansion of the anal lobe of the hindwings took place, since the jugal fold in the hindwing crosses A_1 subbasally (Fig. 8.12a) or basally (Fig. 8.12b–d). In other words, impetus for the modification of the wing structure of Homoptera Auchenorrhyncha and Heteroptera was provided by the need to lock the wings while folded over the back.

Another viewpoint about the venation pattern of the wing anal area of these insects is also possible. An anastomosis between the bases of *PCu* and A_1 (Fig. 8.1, p.133) had already appeared in the Blattinopseida, due to the expansion of the hindwings, with A_2 markedly forked and the jugal fold projecting between A_1 and A_2. This sort of venation is typical of modern Homoptera Auchenorrhyncha, with the richest set of veins and folds in the anal area of the hindwing (Fig. 8.13). On the forewing of *Glaphirophlebia* (Fig. 8.1) the second anal vein, due to longitudinal bending of the clavus during roof-like folding, gives rise to numerous branches which lie outside the jugal fold. Thus, in the Paraneoptera and other neighbouring orders whose representatives possessed expanded hindwings, the

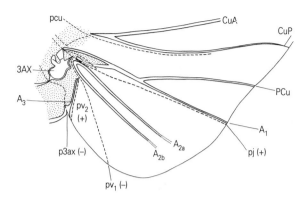

Fig. 8.13 Posterior part of the hindwing base in the froghopper *Aphrophora salicina*.

jugal fold takes a different route in the fore- and hindwings.

Nowhere near all Homoptera Auchenorrhyncha possess expanded hindwings. In cicadas of the superfamily Cicadoidea, the short hindwings are as narrow and rigid as the forewings (Fig. 8.12d). Deformation of the coupled fore- and hindwings in flight is restricted to special hinged articulations on the main veins of the forewing and to the border between the fore- and hindwings. Elastic deformation is only possible along

a narrow band-like strip along the outer margin of the wings, that lacks venation.

Bocharova-Messner (1979) studied the kinematics and deformation of the wings of the cicada *Cicadetta montana* in tethered flight. At the upper point of the trajectory wings of the opposite sides are brought close together so that the angle between them is 15–20° (Fig. 8.14, 1). The wing surface is made taut to the maximum extent at the extreme of the upper stroke, just before the downstroke begins (Fig. 8.14, 1). As it begins the entire forewing deflects slightly downwards, whereas the hindwing retains its original position, which results in a Γ-shaped profile (Fig. 8.14, 2). While maintaining this profile, and with an increased angle of attack, the flapping surface passes the horizontal and approaches the lowest point of the stroke (Fig. 8.14, 3 and 4). As the muscles which raise the wings contract, the distal part of the forewing deflects downwards, along hinged articulations on the veins (Fig. 8.14, 5, 6). Throughout its ascent the flapping surface is twisted and somewhat angular (Fig. 8.14, 7–9). In general, the shape of the flapping surface during both ascending and descending parts of the trajectory is close to that of the hindwings of Polyneoptera at corresponding stages of the trajectory, except that in the cicada, because the wings are rigid, the deformation is

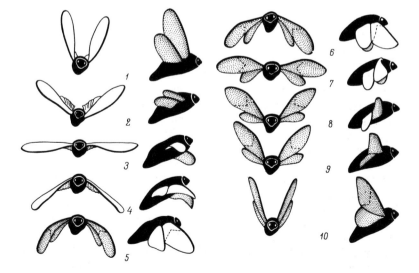

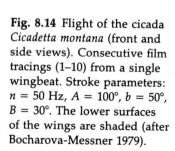

Fig. 8.14 Flight of the cicada *Cicadetta montana* (front and side views). Consecutive film tracings (1–10) from a single wingbeat. Stroke parameters: $n = 50$ Hz, $A = 100°$, $b = 50°$, $B = 30°$. The lower surfaces of the wings are shaded (after Bocharova-Messner 1979).

expressed less explicitly throughout the stroke cycle, and undulating bends are totally absent. The type of deformation of the wings of *Cicadetta montana* during the upstroke should be considered as greatly modified supinational bending, with strong tip deflection. Longitudinal bending takes place along the border between the fore- and hindwing.

8.3.3 The Heteroptera

All the tendencies and changes in the structure and functioning of the wing apparatus typical of the Homoptera Auchenorrhyncha also occur in the Heteroptera, in some cases developed to an even greater extent. In the Heteroptera the dorsal oblique muscle is not as powerful as in the cicadiform Homoptera, but it is similarly shifted forward. Fusion of the anterior and median notal wing processes is more complete in bugs; and furthermore, the posterior notal wing process is reduced. One of the principal ways that the structure of the bug pterothorax differs from that of the cicadiform Homoptera is that in bugs the posterior border of the pronotum and hence the anterior edge of the interalar triangle are situated in front of the scutoscutellar suture. Consequently, only the most lateral parts of the oblique sutures are retained and the fossae are located behind the anterior edge of the interalar triangle. The interalar triangle includes the whole scutellum (excluding its edges) and the posterior

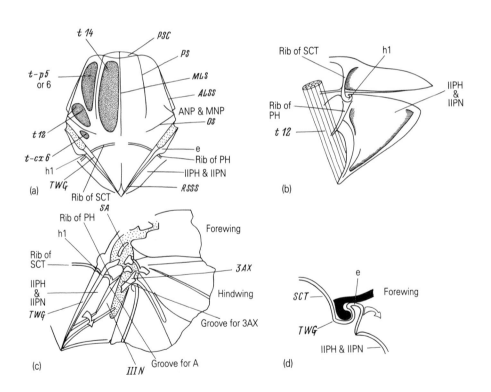

Fig. 8.15 Pterothorax of the plant bug *Adelphocoris triannulatus*, family Miridae: (a) dorsal view; (b) lateral and inner view of the posterior part of the notum; (c) dorsal view of enlarged hindwing base; (d) section through the tergal wing groove. Arrows indicate the direction of displacement of the wall of the interalar triangle as a result of the contraction of the dorsal oblique muscle.

part of the scutum, bordered from anteriad by the edge of the pronotum. The upper ends of muscle *t-cx6* are inserted outside the oblique sutures.

In bugs, there is further refinement of the mechanism for locking the wings when at rest and releasing them before flight takes place. In most bugs the wings are folded flat over the back, and not only the forewings but also the hindwings are locked. In the plant bug *Adelphocoris triannulatus* the hindwing bends along the jugal fold; the edge of the unfolded part of the wing settles flat on the back. The third axillary sclerite sits in a corresponding depression between the metanotum and the wing base, while the anal vein fits into a slit between the mesopostnotum and the metanotum (Fig. 8.15c). The inner edge of the wing is inserted under the free apex of the interalar triangle. To lock the forewing, the bent posterior edge is put under the inclined inward edge of the scutellum (Fig. 8.15d). The anal lobe of the forewing rests on the platform of scutum outside the oblique suture (Fig. 8.15a) and is covered by the pronotum. Clearly the tergal wing groove performs the main locking function: its external wall, which is turned outwards elastically as the wing is released, acts like a push-button (Fig. 8.15d). The edge of the scutellum is turned outwards by the phragma, which in bugs is fused with the postnotum. On the anterior edge of the phragma there is a rib which hinges with the rib on the internal side of the scutellum by a mobile articulation (Fig. 8.15b). Because of the elastic articulation in the main hinge area, contraction of the *t14* muscle, the posterior end of which occupies two thirds of the phragma, causes rotation of the posterosuperior angle of the phragma forwards, inwards, and downwards, while contraction of the dorsal oblique muscle, by contrast, causes rotation of the same angle backwards, outwards and downwards (Fig. 8.15c). Hence, contraction of any of the dorsal muscles causes the wall of the interalar triangle to turn downwards, which is necessary for 'unbuttoning' to release the forewing. Larsen (1949) noticed a long time ago that releasing the wings from their catch on the thorax is brought about by contraction of the *t14* and *t–p5* or *6* muscles. Barber and

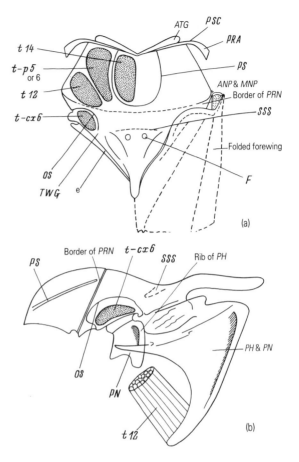

Fig. 8.16 Pterothorax of the damsel bug *Himacerus apterus*, family Nabidae: (a) dorsal view; (b) sagittal cross-section of mesothorax with most muscles removed.

Pringle (1966) showed experimentally that in the Belostomatidae, wing release is powered by the *t12* muscle while *t14* and *t–p5* or *6* enable the released wings to open.

In bugs, as well as in the cicadiform Homoptera, the recurrent scutoscutellar suture is hypertrophied, swollen, and forms a peak beneath which the folded wing is inserted (Fig. 8.16a). The posterior part of the scutum and the scutellum are modified, flattened, and adapted for locking the flat-folded wings. The structure of the interalar triangle is more monotypic in

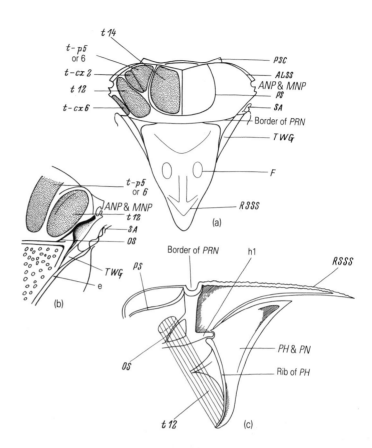

Fig. 8.17 Pterothorax of the stink bug *Dolycoris baccarum*, family Pentatomidae: (a) dorsal view; (b) enlarged region of the notal wing process; (c) sagittal cross-section of mesothorax with most muscles removed.

bugs than in cicadas. The scutoscutellar suture is retained in most cases. In plant bugs, a rib which is clearly visible from above (Fig. 8.15a), corresponds to this suture. In the short-winged bug *Himacerus apterus* of the family Nabidae the scutoscutellar suture is retained as a small groove marked by a pair of fossae behind it (Fig. 8.16a). Furthermore in Nabidae the oblique suture borders a vast area on which the anal lobe of the forewing rests, and to which the well-developed *t-cx6* muscle is attached from the inside (Fig. 8.16b). In short-winged nabids the hinged articulation between the scutellum and the phragma which is necessary for the locking and releasing of the wings is lost, but a rudiment of it remains, located opposite a small transverse depression of the scutoscutellar suture (Fig. 8.16b).

The scutellum structure is most modified in the Pentatomidae, but even these have structures that correspond to fossae and the recurrent scutoscutellar suture (Fig. 8.17a). Nevertheless, the rib of the scutellum is absent, and the rib of the phragma articulates with the caudal outgrowth of a ridge at the pronotum border by a hinge (Fig. 8.17c). In stink bugs there is secondary splitting of the lobe produced by the fused anterior and median notal wing processes, thereby resulting in the notal lobe articulation with the first axillary sclerite acquiring a complex shape (Fig. 8.17b). The cavity which exists between the newly developed processes has nothing in common with the tergal fissure so that the suture that travels backwards obliquely to the centre, contrary to Matsuda's opinion (1970), cannot be homologous with the transscutal sulcus.

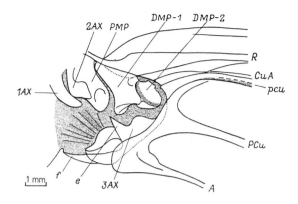

Fig. 8.18 Fragment of the axillary apparatus of the forewing of the bug *Belostoma apache*. The clavus has been slightly omitted to make the underlying structures visible.

We may conclude that the principal momentum in the development of the pterothorax in bugs was specialization for reliable locking of flat-folded wings. The role of the main hinge between the rib of the scutellum and the rib of the phragma is not, however, limited to locking and unlocking of the wings. In the stink bug, the light pressure on the apex of the interalar triangle (Fig. 8.15a and c) causes the parts of the notum near the fused wing processes to displace forwards and outwards, accompanied by abrupt depression and pronation of the forewings. That this movement is possible shows that effort from contraction of muscle *t14* is transferred onto the mesonotum through the middle, rather than through the anterior phragma. In other words, the wing apparatus of bugs has undergone substantial changes: the hinge *h1*, which enables the notum to pivot about the middle phragma, is well developed; and contraction of the dorsal longitudinal musculature is transferred to the notal wing processes via the middle phragma. An analogous situation occurs in the Hymenoptera and Diptera. In all these insects effort resulting from contraction of the dorsal longitudinal muscles is also transferred to the wing bases through the middle phragma. However, bugs are markedly distinct in that the principal momentum in the evolution of their

wing apparatus was the improvement of the system for locking and unlocking the folded wings, rather than for an improvement in the flight system, as in the Hymenoptera and Diptera.

The mechanism by which the wings are folded flat in bugs can easily be derived from that of cicadas, in which the wings are folded in a roof-like position. In bugs, the distal arm of the third axillary sclerite does not overturn at all as muscle *t–p13* contracts, but only moves backwards and in towards the body (Fig. 8.18). *DMP–2* rotates around *DMP–1* with its distal edge downward like half of a door hinge. *R* moves backwards but is not depressed as in cicadas, resulting in flat folding of the wing, and the outgrowths of the sclerites enter into matching fissures, like elements of a mosaic. The hindwing is folded in the same manner. In some water bugs (Notonectidae, Corixidae), roof-like folding of the forewings has appeared secondarily via the partial reduction of *DMP–1* and a decrease in mobility in the hinge between it and *DMP–2* (Betts 1986*a*).

While in flight, the wings of one side are coupled and act as one flapping surface. During the downstroke the hindwings of the bug *Piezodorus lituratus* form a canopy (Fig. 8.19, 5–8). At the end of the downstroke the forewing, along the border of the coreum, bends ventrally, with the tip facing downwards and forwards (Fig. 8.19, 12–16); and when the tip returns to its initial position, the wing pair on one side is bent longitudinally (Fig. 8.19, 14–20). The line of bending follows the border between the fore- and hindwing. Deformation of the hindwings during the upstroke resembles that of the vannus of a stonefly hindwing (see Fig. 7.4, p.119). The difference is that the hindwings of bugs are not folded in a fan-shape at the end of the downstroke, but lag behind the forewings. When they begin to go upwards, a torsional wave passes over their surface from tip to base (Fig. 8.19, 16–24), supinating the entire hindwing.

The analysis of bug wing deformation during flight enables us to reconstruct the functional pattern of the wing pair on one side (Fig. 8.20a). Wing tip deflection occurs along the nodal line (transverse fold) that lies close to the coreum—a sclerotized proximal part of the

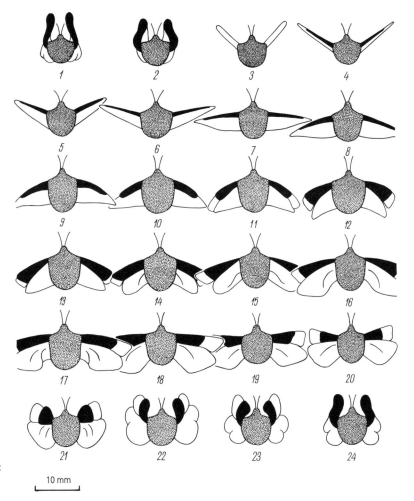

Fig. 8.19 Flight of stink bug *Piezodorus lituratus* (front view): consecutive film tracings (1–24) from a single wingbeat. Stroke parameters: $n = 90°$, $A = 130°$, $b = 30°$, $B = 60°$.

10 mm

remigium. In some species a small fissure passes longitudinally through the centre of the coreum. According to Betts (1986b), the coreum bends a little along this fissure during the stroke, forming a convex wing profile. A fold behind R on the hindwing (Fig. 8.20b) matches this fissure, presumably the fold of the radial sector. In flight, the clavus of the forewing overlaps the radio-medial field of the hindwing; the bending line of the wing pair during supination includes *pcu* of the forewing and the border between the fore- and hindwing (Fig. 8.20a). A characteristic feature of the hindwings of bugs is the forking of *pcu* (Fig. 8.20b), which has caused the flexible, deformable area through which the wave passes during supination to become still more expanded. Additional flexibility and malleability of the deformable zone is also provided by the special structure of the anal veins (Betts 1986a).

8.4 Conclusion

The transition from functionally four-winged to functionally two-winged flight was an important event in the evolution of winged insects, associated with an increase in wingbeat frequency and the abrupt expansion of flight technique possibilities. The mechanism for locking

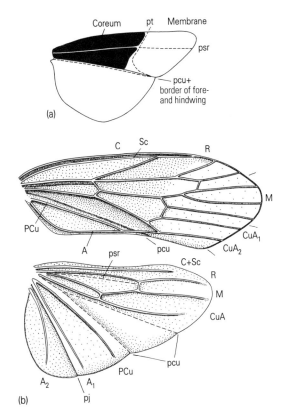

Fig. 8.20 Functional wing design in the Heteroptera. (a) The organization of the fore- and hindwings as a flapping surface (adapted from Bocharova-Messner 1982). (b) Hypothetical ground plan for heteropteran wings (after Wootton and Betts 1986, with lettering changed).

the wings in a roof-like position was improving at the same time. Compared with flat folding, this type of folding requires specific mechanisms for reliable locking. The first synapomorphy of functionally four-winged insects, the appearance of muscle *t–p13*, was associated with improved locking of the folded wings. The second synapomorphy, linked with flight refinement, was the formation of the scutellum, concentrating effort from notal distortion on to the anterior and median wing processes. In the common ancestor of the Para- and Oligoneoptera, as well as in the Hypoperlida, the fore- and hindwings were equally developed, save that the hindwings were slightly shorter and sometimes had an expanded anal lobe. The wing kinematics of the insects of ancestral group seems to have been close to the functionally four-winged type that persisted in the Neuroptera, Megaloptera, Mecoptera, primitive Trichoptera, and in early moths with uncoupled wing pairs. Deformation of the wings occured by supinational bending and occasionally, as in primitive caddisflies, the wings underwent supinational twisting during the upstroke.

Later, in the Paraneoptera, changes occurred to do with both flight refinement and an increase in the reliability of wing locking. Flight improvement influenced the fusion of the anterior and median notal wing processes, and later the reduction of the posterior process; and also hypertrophy of the dorsal oblique muscle. The main emphasis during the reconstruction of the wing apparatus was on locking the folded wings in a roof-like position. Tergal wing grooves developed at the posterior part of the scutum. The forewings settled into these grooves when at rest and were locked using a special device. In the axillary apparatus joint B became reduced, and mobility of the proximal medial plate was lost.

9

Progress in insect flight

I do not know whether I was then a man dreaming I was a butterfly, or whether I am now a butterfly dreaming I am a man.

Chuang Tzu

In the evolution of the Oligoneoptera, the tendency towards improved flight predominated over the tendency to improve the locking of the folded wings. With the complete metamorphosis characteristic of these insects, the function of feeding is accomplished by a poorly mobile larva, while winged imagos retain the functions of the search for a sexual partner and reproduction. The history of the Oligoneoptera began with the order Palaeomanteida (= Miomoptera), known from the Late Carboniferous until the Early or Middle Jurassic. The Palaeomanteida closely resembled the Hypoperlida in their general appearance and wing venation. On the other hand, certain features of the Palaeomanteida were similar to Blattinopseida—the latter had similarly stoutish bodies and a widened hindwing base.

The wings were folded in a roof-like position over the abdomen in the Palaeomanteida; the clavus, like that of Blattinopseida, bent longitudinally, with 4–5 parallel branches of A_1 and A_2. In contrast to those of the Blattinopseida, the jugal folds ran in the same way both on the fore- and hindwings of the Palaeomanteida, between A_2 and A_3. This peculiarity was characteristic of all lower Oligoneoptera, which emphasizes the difference between the ways a roof-like wing folding position was achieved in the Para- and Oligoneoptera. The descendants of the Palaeomanteida can be divided into two distinct groups, to one of which the Hymenopteroidea belong, while the other includes the Coleopteroidea, Neuropteroidea, and Mecopteroidea.

9.1 Improvement of the wing movement mechanism

The most significant changes in oligoneopteran flight apparatus morphology were connected with transformation of the phase relations in the movement of the wing pairs. The functionally four-winged flight characteristic of the Palaeomanteida was retained in the lower Coleopteroidea and Mecopteroidea and in all Neuropteroidea. In the Hymenopteroidea and higher Mecopteroidea there was preferential development of the forewings, and the wing musculature was concentrated in the mesothorax. In the Megaloptera, apart from the increase in the stroke amplitude of the hindwings, the

base of the hindwing became broadened. By a similar process, the forewings of the ancestors of beetles seem to have gradually taken on the function of an integument, which eventually led to the appearance of true hind-motor insects—the Coleoptera.

With the exception of rose chafers (*Cetonia*), which keep their elytra (forewings) closed, all Coleoptera spread their elytra, usually at a dihedral angle, but most do not keep them still. Many beetles swing them at low amplitude at the same frequency as the hindwings (approximately 50 Hz). Elytra may work as oscillating 'guiding vanes' (similar to stators in turbines) to improve airflow to the hindwings. The obliquely aligned body of *Melolontha* generates a small amount of lift, too.

Tiger beetles (*Cicindela*) spread their elytra and wings very quickly and take off rapidly without the complicated warming-up preparations typical of many beetles. The elytra are stretched more or less horizontally and are supported by the forelegs like the struts of a high-wing monoplane. Coleopteran flight has been classified kinematically with respect to functional morphology by Schneider (1982).

9.1.1 The skeleton and musculature

Several changes are needed to get from the generalized scheme of the wing-bearing tergal plate (see Fig. 1.5, p.10) to a scheme that takes into account the most general features of oligoneopteran tergum organization. First of all, the side branches of the recurrent scutoscutellar suture must have become reduced so that the scutoscutellar suture became significantly longer than the recurrent one, and the main hinges, which are known to have been situated at the point where both sutures cross, appear to have been at the posterior border of the notum. Second, the lateropostnota spread over the dorsal surface of the wing, frequently fusing with the posterior notal wing processes. Third, the oblique sutures on either side of the notum reached as far as the notal border, thus separating the median notal process from the posteromedian one. And finally, in none of the oligoneopterans are the lateral parapsidal sutures developed. The sutures described as such in the Hymenoptera and Diptera (Matsuda 1970) are not homologous with the lateral parapsidal ones, as will be demonstrated below. If we take into account all the changes mentioned, then a generalized scheme of the wing-bearing tergal plate of the Oligoneoptera should look like that shown in Fig. 9.1.

Megaloptera (see Fig. 1.9, p.14). In these insects, as in other Oligoneoptera, there is a true scutellum which is furnished with the paired muscle *t13*. The scutoscutellar suture is significantly longer than the recurrent scutoscutellar suture.

The main hinges are situated where these two sutures cross, from which the oblique sutures run towards the notal border. The anterior end of muscle *t13* is attached to the corresponding scutoscutellar branch all along its length, while on the other side, muscle *t12* is fastened to the same branch in a similar manner. Muscle *t–cx6* is well developed, and its upper end is attached to the notum at a rather typical place—opposite the posterior notal wing process. The other dorsoventral muscles are attached in an integrated morphofunctional bundle opposite the median notal process, which bears the main load during wing operation.

Coleoptera (Fig. 9.2). In the beetle *Potosia metallica*, as in most other beetles (Matsuda 1970), the scutum is divided by a transverse fissure into an anterior membranous part and a posterior sclerotized part. Outwardly, the fissure resembles the transscutal sulcus of the Orthoptera; however, unlike the latter, the fissure runs into the notal border directly in front of the anterior notal wing process (Fig. 9.2). In the places where the anterior membranous part of the scutum is overhung by the posterior sclerotized shield, its borders bear strong ribs called alacristae. Each alacrista is crowned by a capitate apodeme which forms

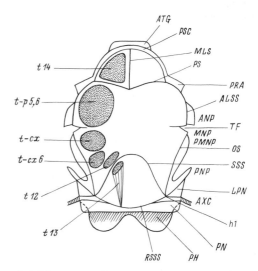

Fig. 9.1 The main features of the tergum of a wing-bearing segment in the Oligoneoptera.

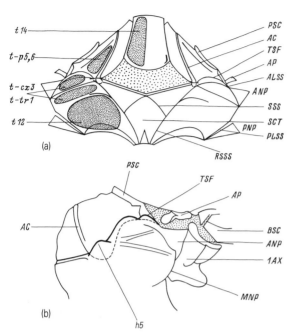

Fig. 9.2 Metatergum of the beetle *Potosia metallica*: (a) dorsal view and (b) enlarged region of the anterior notal wing process.

a mobile joint, a kind of a hinge (*h5*), with the posterior sclerotized shield. During contraction of the dorsal longitudinal muscles both parts of the notum bend along the transverse fissure; the posterior part rotates backwards relative to the apodemes of the alacristae, lifting the anterior notal wing processes slightly. The muscle *t–p5, 6* is attached to the notum on the outside of the alacrista. The tergocoxal muscles and the muscle *t–tr1* are fastened to the notum in two small bundles, opposite the anterior wing process. The clearly secondary absence of muscle *t13*, which became reduced as a result of significant transformations in the entire tergal region, as well as the vigorous development of muscle *t12*, can be added to the list of skeleto-muscular characteristics of beetles. Thus, similar changes in the metathorax of both the Coleoptera and Orthoptera have been associated with the hind-motor state. Further development of the longitudinal dorsal muscles is limited in the

hind-motor state, so that most of the work of lowering the wings falls on the basalar musculature. This in turn implies that wing operation by the indirect flight muscles must occur via the anterior notal wing process. It is this that is most developed in the Coleoptera and the Orthoptera. The changes in the wing apparatus organization of the Coleoptera (as compared with the Megaloptera) and the Orthoptera (as compared with the Plecoptera) can be considered convergent, with a net outcome, from a mechanical point of view, of wing operation via the anterior notal wing process. However, this result is achieved in different ways, by means of structures that are not homologous—the transscutal sulcus in the Orthoptera, and the transscutal fissure in the Coleoptera.

Neuroptera (Fig. 9.3). First of all, it should be noted that the neuropteran mesonotum is of a generally primitive structure. This manifests itself in the different muscle groups being of approximately the same size. Both tergopleural muscles, *t–p5* and *t–p6*, the group of tergocoxal muscles, and the dorsal oblique muscle form four bundles of equal volumes. Muscles *t13* and *t–cx6* are relatively strongly developed. The lateropostnotum and the posterior notal wing process are fused together to form a functionally integrated join with the third axillary sclerite.

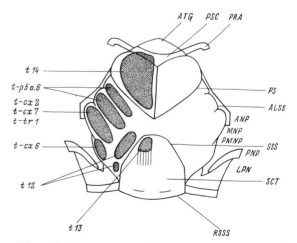

Fig. 9.3 Mesotergum of the antlion *Acanthaclisis occitanica*, dorsal view.

Mecopteroidea (Figs 9.4 and 9.5). The main feature of the pterothorax of the Mecopteroidea (scorpionflies, flies, caddisflies, butterflies, and moths) is the lack of muscle *t–cx6*. Mickoleit (1966) considered this feature to be synapomorphic for the mecopteroideans. In addition, the tergum of these insects is characterized by the presence of a well-pronounced scutellum to which muscle *t13* is attached. In flies this muscle has been lost secondarily. The shape of the scutellum tends to vary depending on how muscles *t13* and *t12* are fastened to the scutoscutellar suture. Thus, the mecopteran scutellum is short

and flat-topped (Fig. 9.4a), while in caddisflies it is elongate and has a sharpened top (Fig. 9.4b). In the Mecoptera the muscles, which are functionally homogeneous as regards wing movement, form a morphofunctional bundle attached to the notum opposite the anterior and median wing processes (Fig. 9.4a). The parapsidal sutures are not developed in these insects, as the attachment region of the anterior end of muscle *t14* does not overlap the scutum.

Beginning from the characteristic wing apparatus of scorpionflies, it is easy to follow the changes which led to the formation of the dipterous tergum (Fig. 9.5). The pterothorax of the higher Diptera is the biomechanical acme of the system of wing movement. To begin with, the dipterous tergum is characterized by great mobility of the scutellar arms, which rotate around hinges which are homologous to the principal hinge (Fig. 9.5a (ii) and (iii)). The scutellar arm bears the posteromedian notal wing process. In addition, one more lever has appeared, formed by fusion of the posterior wing process with the lateropostnotum (Fig. 9.5a(ii)). Finally, as the set of muscles became reduced the tergotrochanteral muscle increased in volume, and the muscle *t–cx7* split into two. Numerous authors (Nachtigall 1969, 1978; Anderson *et al.* 1985; Schouest *et al.* 1986) have shown how muscle *t–tr1* acts as a starter motor in dipterans, by allowing the insect to make the initial leap and causing simultaneous deformation of the notum. The contraction of the antagonistic dorsal longitudinal and dorsoventral muscles is triggered at the same moment. The increase in volume of muscles *t–tr1* and *t–cx7*, which are important for flying and walking, resulted in the appearance of additional, transnotal, sutures in dipterans. Despite the outward similarity between these and the lateral parapsidal sutures, formation of the transnotal suture is not associated with muscle *t–p5, 6* but with the increase in the volume of the tergotrochanteral muscle due to the characteristic behaviour of the Diptera, involving frequent take-offs and landings. The transnotal sutures are well developed in *Rhagio* (Fig. 9.5a (i)), while in *Sarcophaga* (Fig. 9.5b) they join the parapsidal sutures medially then cross the notum.

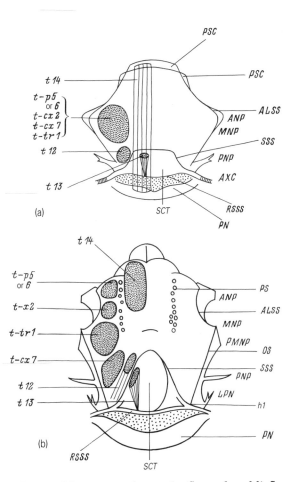

Fig. 9.4 Mesoterga of scorpionfly and caddisfly, dorsal view. (a) Scorpionfly *Panorpa* sp. and (b) caddisfly *Phryganea bipunctata*.

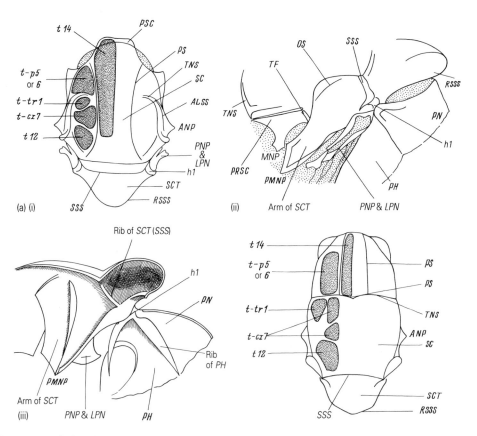

Fig. 9.5 Structure of the dipteran mesothorax. (a) Snipe fly *Rhagio scolopaceus*: (i) dorsal view of the mesotergum and posterior part of the notum, seen from (ii) the outside, and (iii) the inside, the head to the left. (b) Carrion fly, *Sarcophaga* sp., dorsal view of the mesotergum.

The mecopteran scutellum is able to move to some extent relative to the scutum. This peculiarity seems somehow to have determined the role of the scutellum in dipteran wing movement. The reduction in muscle *t13* was reflected in a dramatic increase in the mobility of the scutellum. Improvement of the wing operation mechanism had the effect that the scutoscutellar suture no longer functioned as a support for muscle *t12* in Diptera, while the scutellum acquired its typical shape, which is associated with the intricate movement mechanics of its arms (Ennos 1987; Brodsky 1988, 1989, etc.); in particular, the recurrent scutoscutellar suture is better developed than the scutoscutellar one.

In the Trichoptera and Lepidoptera, muscles

t–tr1 and *t–cx7* are also characteristically strongly developed (Fig. 9.4b); however in these insects *t–cx7* is displaced backwards to occupy the position of the lacking *t–cx6* muscle, which meant that the general notal structure did not undergo any significant changes, and there are no additional sutures. The shape of the scutoscutellar suture was altered by the backward shifting of the upper end of muscle *t–cx7* and the poor development of *t12*. The caddisflies are characterized by having approximately equal muscle volume in the various muscle groups, which can be regarded as a primitive feature. In these insects muscle *t–p5* or *6* is fastened to the ventral side to the basisternum. Furthermore, the Trichoptera (Ivanov and Kozlov 1987) and lower Lepidoptera (Kozlov 1986) retain muscles *t–cx2* and *t13*, which are lacking in the Diptera.

The Mecopteroidea exemplify how in the course of evolution the separate notal regions lose their corresponding muscles, the different muscles are integrated into morphofunctional bundles, and how the deformability of the liberated notal regions increases. Thus the muscle ends on the inner surface of the notum of the caddisfly *Phryganea bipunctata* (Fig. 9.4b) are arranged in metameric order, filling almost the entire notal surface. By contrast, in the fly *Sarcophaga* sp. (Fig. 9.5b), significant regions of the scutum are free of muscles; one such region is separated from the others and, at the level of the median wing process, forms the parascutum (Fig.9.5a (ii)) and participates in control of wing movement (Boettiger and Furshpan 1952; Miyan and Ewing 1985, etc.).

Hymenoptera (Fig.9.6). In these insects the process of reduction of the dorsoventral muscles is far more advanced than in any other insect group. The Vespidae retain only one pair of dorsoventral muscles—*t–p5 or 6* (Fig. 9.6c). The decrease in number of muscle bundles was accompanied by an increase in the volume of the remaining ones. A crista, which projects deeply into the body cavity, has formed on the sternum on which the extremely massive muscle *t–p5 or 6* is fastened. The parapsidal sutures have developed progressively in association with elongation of the scutum: in the Symphyta these sutures are short and have fused together along the median line (Fig. 9.6a), whereas in the Apocrita they are long and free (Fig. 9.6b and c). In addition, in higher hymenopterans the strength of the attachment of the anterior ends of the longitudinal dorsal muscle is increased by the well-developed medial longitudinal suture, which has a correspondingly deep keel inside. Finally, in the higher Hymenoptera a pair of new sutures have appeared, but not induced by the attachment of the muscles to the inner notal surface. The supplementary parapsidal sutures run parallel to each other and to the parapsidal sutures, from the anterolateral angles of the notum to the crossing point of the transscutal sulcus and the scutoscutellar suture (Fig. 9.6c). Development of the supplementary parapsidal

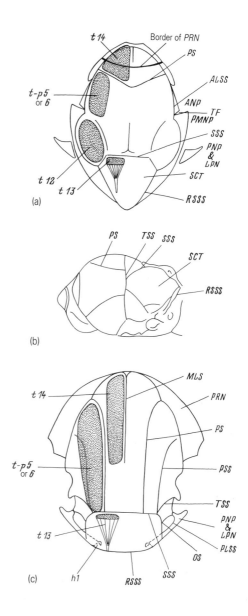

Fig. 9.6 Structure of the hymenopteran mesothorax. (a) The horntail *Urocerus gigas*, dorsal view of the mesotergum. (b) The chalcid *Brasema* sp., laterodorsal view of the mesotergum, head to the left (after Gibson 1986). (c) The common wasp *Vespa orientalis*, dorsal view of the mesotergum. (a), (b), and (c) are not to scale.

sutures was probably associated with the change in the nature of notum deformation which was induced by contraction of the indirect flight muscles.

In the Hymenoptera, there is also a well-developed transscutal sulcus in the tergum, which performs a significant role in wing movement. The transscutal sulcus is not present in the Symphyta, but the tergal fissures are well-pronounced (Fig. 9.6a). In the Parasitica, the transscutal sulcus traverses the notum (Fig. 9.6b), while in the Aculeata it follows the scutellar angle on either side of the notum, delimiting the scutellar arm anteriorly (Fig. 9.6c). Consequently in Hymenoptera, unlike in Diptera, this arm bears not only the posteromedian notal wing process but also the median one. The scutellar arms move around the hinges which are homologous to the main one. An additional lever is formed by the fusion of the posterior notal wing process and the lateropostnotum (Fig. 9.6c). Thus in the Hymenoptera and Diptera there were parallel changes in tergal structure which are associated with increased scutellar arm mobility and a more complex wing operating mechanism.

9.1.2 The axillary apparatus

Transfer of the centre of gravity from the anterior notal to the median wing process during wing operation is associated with significant transformation of the axillary apparatus. First of all, differentiation of the notal margin occurred (Fig. 9.7).

In the Megaloptera and Coleoptera the typical shape of the first axillary sclerite is preserved, but the angle *c* is elongate and sharp-pointed (Fig. 9.7, d–f). The reduction of the contact area between the first axillary sclerite and the notal margin in the median wing process region resulted in the formation of an additional contact between these two structures, and hence the posteromedian wing process was formed. In the Corydalidae this process has developed on the forewings only. The posteromedian wing process is absent from the hindwings of leaf beetles (Fig. 9.7e), but is well developed in ground beetles (Fig. 9.7f). In addition, the axillary apparatus of the Coleoptera is characterized by

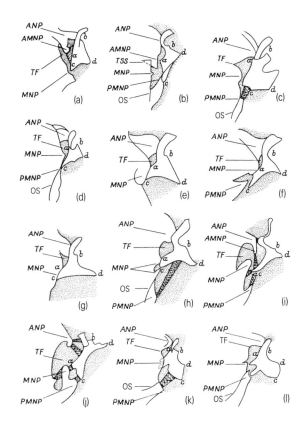

Fig. 9.7 Structure of the notal margin and the first axillary sclerite in various insects: (a) grasshopper *Anacridium aegyptium* (after La Greca 1947), (b) horntail *Urocerus gigas*, (c) antlion *Myrmecaelurus trigrammus*, (d) dobsonfly *Corydalis* sp., (e) leaf beetle *Zygogramma suturalis*, (f) ground beetle *Calosoma sycophanta* (after Tietze 1963), (g) scorpionfly *Panorpa communis*, (h) hepialid moth *Hepialus humuli*, (i) nepticulid moth *Stigmella basalella* (after Sharplin 1963a), (j) owlet moth *Plusia gamma* (after Sharplin, 1963a), (k) crane fly *Phalacrocera replicata*, (l) hover fly *Helophilus affinis*. (a)–(d), (g), (i), and (j) forewing; (e), (f), and (h) hindwing.

a close functional contact between the second axillary sclerite and the basiradiale, as a result of which the anterior margin of the second axillary sclerite has developed a curious platform (Fig. 9.8). The presence of an especially close contact between the tip of the arm of the first axillary sclerite and the basisubcostale suggests that the basalar muscles play a significant role in wing movement.

In the Neuroptera, the most striking feature of the axillary apparatus is the formation of a complex joint between the first and second axillary sclerite, that is visible in the complicated outline of the first axillary sclerite (Fig. 9.7c). There is a well pronounced bond, by means of a copula, between the posteromedian notal process and the cut angle *c* of the first axillary sclerite. There is no direct contact between the first axillary sclerite and the posteromedian notal wing process in these insects.

In the Mecopteroidea, the border of the first axillary sclerite *a–c* is reduced, and the border *c–d* is widened, which can be easily seen

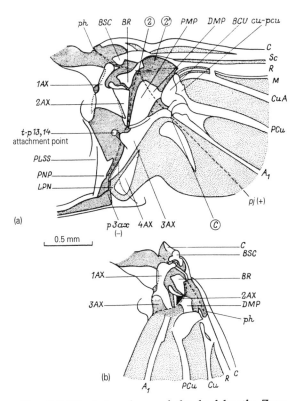

(a)
0.5 mm

(b)

Fig. 9.8 Hindwing base of the leaf beetle *Zygogramma suturalis*. Dorsal view of arrangement of the sclerites on (a) the extended and (b) the folded wing.

in the scorpionfly (Fig. 9.7g). As a result, the short wing lever is longer, leading directly to greater flap efficiency. In the hindwings of the Lepidoptera the median notal wing process is split by a short fissure into two processes (Fig. 9.7h), only the posterior of which is retained in the forewings; and the fissure referred to above unites with the tergal fissure. As a result, the tergal fissure assumes its characteristic shape and its oblique posteromedian orientation (Fig. 9.7i). The shape of the fissure means that contraction of the dorsal longitudinal musculature in Lepidoptera and Trichoptera leads not only to a lifting but also to a shifting forward of the median wing process. Furthermore, narrowing of the border *a–c* is accompanied by the development of the posteromedian wing process, which is attached by a copula to the middle of the border *c–d* (Fig. 9.7h). In more advanced Lepidoptera strengthening of the posteromedian wing process is apparent (Fig. 9.7i). In addition, the anterior wing process turns backward to form a further area of contact with the first axillary sclerite (Fig. 9.7i). In the Lepidoptera, transfer of the centre of operation of the first axillary sclerite to the posteromedian notal wing process is associated with division of the arm of the first axillary sclerite (Fig. 9.7i). In higher Lepidoptera, for instance in the Noctuidae, the border *a–c* has become widened secondarily, which has eventually resulted in the formation of a quite unusually-shaped first axillary sclerite (Fig. 9.7j).

In the Diptera, the formation of two apodemes on the first axillary slerite, to which muscle *t–p10* is attached by two bundles, has caused widening of the border *a–c* (Fig. 9.7k). The border *c–d* is shifted distally, to join with the third axillary sclerite (Fig. 9.7k and l). In contrast with the Hymenoptera, in which the posteromedian notal wing process forms a direct contact with the first axillary sclerite even in lower representatives of the order (Fig. 9.7b), in Diptera, the consecutive stages of development of the posteromedian notal wing process are as follows: in Nematocera the posteromedian process is connected to the border *a–c* by a copula (Fig. 9.7k), whereas in the Brachycera this process covers the top of the scutellar arm, which is joined to the posterior apodeme of the *a–c* border of the first axillary sclerite (Fig. 9.7l).

In the Hymenoptera, the *c–d* border of the first axillary sclerite has become broader (Fig. 9.7b). The *a–c* border has become narrow, and makes contact with the notal border region which is situated behind the median notal wing process and bounded by an oblique suture. A new functional contact (*PMNP*) with the body of the first axillary sclerite has appeared, at the expense of the development of the median wing process and, in consequence, one further contact region has formed between the notal border and first axillary sclerite—the anteromedian notal wing process (Fig. 9.7b). Both of the areas in which the notal border makes contact with the first axillary sclerite are separated from the posteromedian notal wing process by the tergal fissure, which in these insects is often continued medially in the form of a sulcus.

Thus in all holometabolic insect orders the notal margin is differentiated and specialized. The anteromedian notal wing process is formed independently in the Lepidoptera and Hymenoptera, while the posteromedian notal wing process has developed independently as an additional direct support for the first axillary sclerite in the Hymenoptera, Megaloptera, Coleoptera, and higher Diptera. The furthest extreme is the formation in the higher Hymenoptera and Diptera Brachycera of scutellar arms which possess great mobility and play an important role in wing joint mechanics (Boettiger and Furshpan 1952; Pringle 1968; Miyan and Ewing 1985; Ennos 1987; Pfau 1987).

The anterior region of the notal margin became strengthened in parallel with the weakening of the posterior notal wing process whose base first narrowed and then in some cases separated completely from the notum. As the third axillary sclerite loses its direct contact with the notum, difficulties appear in wing remotion and pronation regulation at the end of the upstroke, and the muscles of the third axillary sclerite, whose operating accuracy is reduced due to the loss of support in joint B, become less efficient.

Weakening of the posterior notal wing process along different evolutionary lines reaches its maximum in the Para- and Oligoneoptera.

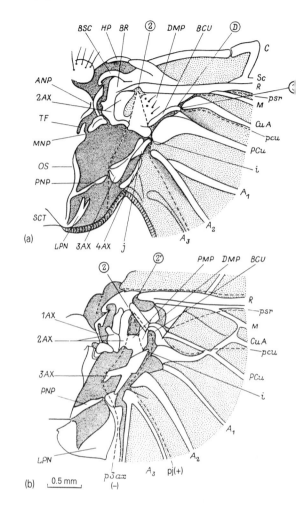

Fig. 9.9 Bases of (a) the fore- and (b) the hindwings of the caddisfly *Phryganea bipunctata*.

The Paraneoptera have less 'problems' with the weakening than do the Oligoneoptera, as the Paraneoptera have retained more contact between the posterior arm of the third axillary sclerite and the subalar sclerite. In connection with this, the changes which compensated for loss of the union of the posterior wing process with the notum only became one of the leading trends in the evolution of the axillary apparatus in the Oligoneoptera. In these insects the lateral postnotal process, which runs through the wing

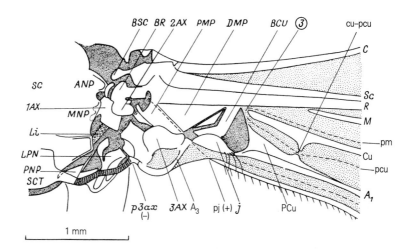

Fig. 9.10 Wing base of the
crane fly *Phalacrocera replicata*.

membrane, comes out on dorsal surface of the
wing and forms a functional contact with the
posterior notal wing process. The basic form of
the lateropostnotum is a characteristic bilobed
shape, as in some caddisflies (Fig. 9.9) and in
Diptera Nematocera (Fig. 9.10). Later on, the
lateropostnotum is fully (Fig. 9.11) or partly
(Fig. 9.12) fused with the posterior wing process;
however, in most cases both these structures
can easily be distinguished on the wing dorsal
surface (Fig. 9.13). In the higher Hymenoptera
and in Diptera Brachycera the fused posterior
wing process and lateropostnotum form a so-
called 'axillary lever,' whose function has been
interpreted in several different ways. According
to Pringle (1968), the axillary lever of the Apidae,
which has muscles attached, stretches the dorsal
longitudinal muscles at the end of the upstroke
and so increases the force of contraction of
these muscles during the downstroke. Miyan
and Ewing (1985) showed that in the Diptera
Cyclorrhapha the function of the axillary lever is
to control the thoracic and alar squamae and the
wing stroke parameters by influencing the position of the third axillary sclerite. The top of the
axillary lever is certainly connected to the third
axillary sclerite (Fig. 9.11) which, in turn, forms a
joint with the first axillary sclerite. Thus, in both

the higher Hymenoptera and higher Diptera the
posterior notal wing process, having fused with
the lateropostnotum, has begun to take an active
part in wing operation.

In a number of cases, for example in the
Hymenoptera Symphyta, Megaloptera, Trichoptera, etc., either the lateropostnotum or its top
has become separated, acquiring independence
and becoming included in a triangle which is
tipped over when the wing folds (Figs 9.9, 9.12,
and others). The top of the lateropostnotum might
be conveniently designated as the 'fourth axillary
sclerite', although this name is usually applied to
the posterior wing process when it is separate.
However, whether the posterior wing process is
separate or not, its nature remains unchanged,
whereas calling it the 'fourth' implies that it is additional to the three pre-existing axillary sclerites,
and that the name is therefore appropriate for the
new structure, the separated lateropostnotal region.
Incidentally, this name was used by Richards (1956)
for this sclerite in the Hymenoptera.

The process of evolution saw frequent liberation of the wing to rotate around its longitudinal
axis, causing reconstruction of the structures of
the distal zone of the axillary apparatus. Performance of the torsional hinge gradually improved.
In the most primitive case (the Megaloptera),

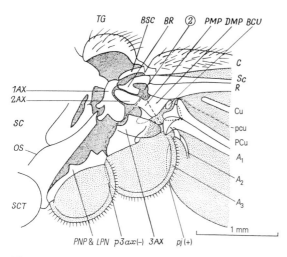

Fig. 9.11 Wing base of the hover fly *Helophilus affinis*.

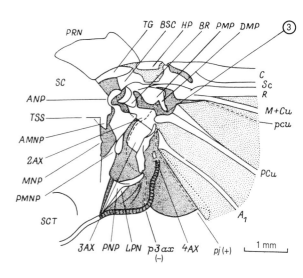

Fig. 9.12 Forewing base of the horntail *Urocerus gigas*.

the bases of the medial and cubital veins are separated from the sclerites by a radial sector fold. This is almost the case in the structure of the torsional hinge in the lower Lepidoptera (Fig. 9.14a). In primitive caddisflies the strong base of the cubital vein forms part of the torsional hinge in the forewings (Fig. 9.9), while in the Neuroptera the base of the cubital vein forms a well-pronounced joint with the basicubitale (Fig. 9.13). The next step in the development of the torsional hinge was the formation of a capitulum joint between the distal medial plate and the base of *PCu*, which is characteristic of the Hymenoptera (Fig. 9.12). The wing's ability to

rotate around its longitudinal axis is increased even more with the formation of a joint between *R* and *A₁* (Fig. 9.15).

In several coleopterans the torsional hinge structure is relatively primitive, and the joint is formed between *R* and *PCu* (Fig. 9.8); in larger forms, the base of *R* itself is twisted during strokes (Pfau and Honomiche 1979), as it is in large caddisflies (Brodsky and Ivanov 1986). The formation of constrictions at the base of *R* is especially typical of Diptera, occurring in all Brachycera independent of their body size. In these insects several torsional hinges are situated together at the wing base (Fig. 9.15).

9.2 Improvements of the wing folding and locking mechanisms

Improvement in the folding and locking of roof-like wings in the Oligoneoptera differed from that in the Paraneoptera, and was associated with the appearance of new structures. Two mechanical systems developed at the wing bases which ensured roof-like folding: a hinge and lever system, and a sliding stop system. The hinge and lever system is usually found on the forewings, and the sliding stop system on the

hindwings. The hindwings of the Corydalidae, as well as the forewings of the Hepialidae (Fig. 9.14), have both these systems.

The hinge and lever system (Fig. 9.16a). During the elevation of the distal arm of the third axillary sclerite, *DMP* rotates in the joint C, bending its apex down; the outer margin of *DMP* influences the mechanic axis of the wing via joint D, causing the folding wing to turn

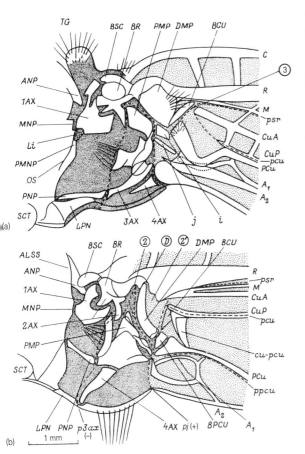

Fig. 9.13 Bases of (a) the fore- and (b) the hindwings of the antlion *Acanthaclisis occitanica*.

body; meanwhile, the anterior end of the inner margin of *DMP* stops against the mechanical axis of the wing at a point (joint D) more proximal than joint C; that is, at a point where the torque is weaker. As a result the wing surface is bent longitudinally along the distal branch of the humeral fold. At the second stage of folding the wing is drawn backwards, rotating around the hinge 2'. The psocopteran wings have a similar sliding stop system; however, in psocids stop is provided by the edge of *PMP* and, additionally, the distal arm of the third axillary sclerite is not turned over but only moves backwards and towards the body.

The sliding stop system is especially well developed in the hindwings of the Coleoptera (Fig. 9.8), Megaloptera (see Fig. 1.13, p.22), Lepidoptera, and Trichoptera (Fig. 9.9). In the latter two groups joint D is formed between *BCU* and the inner margin of *DMP*. The joint between *PMP* and *DMP* on trichopteran hindwings is so elastic that *PMP* is bent still further during wing folding, and thus divided into proximal and distal lobes (Fig. 9.9). Formation of the hinge 2' is associated with membranization of the base of *R*, which is especially distinct in some beetles (Fig. 9.8).

There is also a stop system on the ventral side of the wing joint. During wing folding in alderflies, the inferior layer of the basisubcostale, which rests on the basalar sclerite (see Fig. 1.11, p.19), hinders the lowering and backwards movement of the wing. However, with further contraction of the muscles of the third axillary sclerite, the process of the basisubcostale inferior layer slides along the upper margin of the basalar sclerite, and consequently as the wing turns its costal margin down, it 'clicks' from one fixed position to another. A similar stop system is found on the ventral side of the wing joint in Coleoptera.

Many insects which fold their wings in a roof-like position have devices which allow the folded wings to lock firmly. Such specializations include, first, the presence of both *t–p13* and *t–p14* muscles of the third axillary sclerite and, second, various changes in the structures of the wing bases and of the thoracic regions in which the wing bases are secured. For example, rough areas develop on the wing bases and those tergal

its costal margin downwards. Along the outer margin of *DMP*, between the bases of *Cu* and *PCu*, there is a sclerite (*BCU*), which appears to be part of the cubital vein which has become separated. On the forewings of the Corydalidae (see Fig. 1.13, p.22), Trichoptera (Fig. 9.9, p.160), and Lepidoptera (Fig. 9.14), *BCU* forms joint D with *DMP*. In the Neuroptera (Fig. 9.13), *BCU* is especially well developed on the forewings and forms joint D directly with *R*.

The sliding stop system (Fig. 9.16b). As the distal arm of the third axillary sclerite rises and the wing rotates backwards around the hinge 2, the inner margin of *DMP* is also drawn towards the

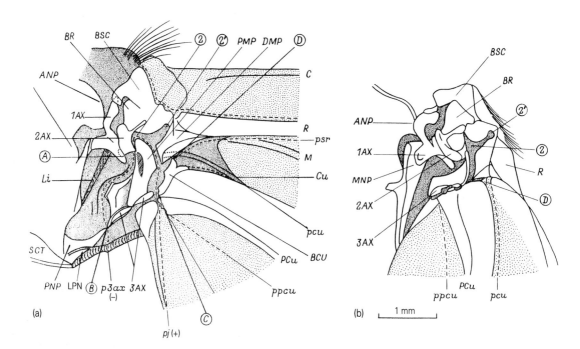

Fig. 9.14 Forewing base of the hepialid moth *Hepialus humuli*. Dorsal view of the arrangement of the sclerites on the (a) extended and (b) folded wing.

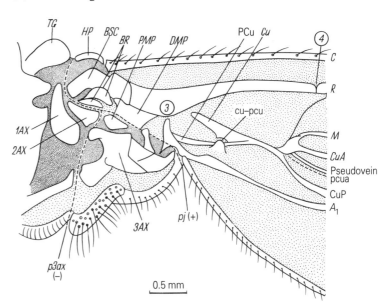

Fig. 9.15 Wing base of the robber fly *Neoitamus cothurnatus*. The circled '4' indicates an additional torsional hinge, located at the base of *R*.

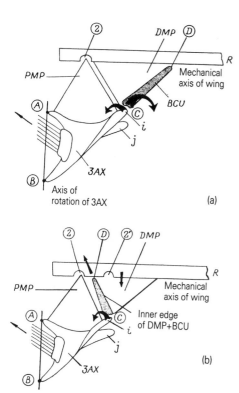

Fig. 9.16 The mode of action of the device allowing roof-like wing folding in the Oligoneoptera. Solid arrows indicate the direction of movement of the sclerites during wing folding, and the thin arrow shows the contraction of the *3AX* muscles. (a) Hinge and lever system characteristic of the forewing. D is a joint between *BCU* and the mechanical axis of the wing. (b) Sliding stop system characteristic of the hindwing. 2' is an additional vertical hinge enabling roof-like folding of the hindwing, as explained in the text.

regions that are in contact with the folded wings. In the Neuroptera, the wing posterior margin is thickened and gets sheathed in a depression on the mesotergum. The shape of the curve of the base of A_2 on trichopteran forewings matches the shape of a depression between the posterolateral parts of the scutum and the scutellum of the mesotergum (Fig. 9.17); the wings of these insects overlap when folded. In the lower

Lepidoptera, whose wings meet with their rear edges but do not touch the mesotergum, the anal veins of the forewing do not take part in locking and are reduced (Fig. 9.14).

The apparatus that locks the folded wings is often retained in insects which fold the wings flat, such as the Meropidae (Hlaváč 1974) and the Hymenoptera Symphyta (Schrott 1986).

A perfect locking apparatus for folded wings occurs in the Diptera Cyclorrhapha. The apex of the pleural wing process is swollen into a head and bears three incisions of different shapes and depths (Fig. 9.18a,b, and c); as the inferior layer of the basisubcostale projects into the anterior incision (a) the longitudinal axis of the wing is drawn noticeably backwards (1). When the basisubcostale is locked in the next incision (b), the longitudinal axis becomes more vertical (2). The posterior incision (c) is the deepest: when the basisubcostale is locked in this incision, the wing is fully folded (3). The folded wing is therefore stabilized when the basisubcostale is locked in the posterior incision, and consequently the bifunctional wing muscles, having switched over completely to operation of the legs, allow the insect to run fast. The basisubcostale can only be shifted from one incision to another when the wing is fully raised, and then the tooth of the basisubcostale stops against the point in front of the posterior incision. With further contraction

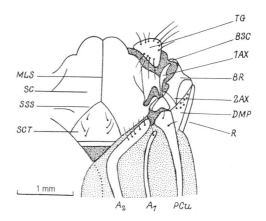

Fig. 9.17 Base of the right forewing of the caddisfly *Phryganea bipunctata*. The forewings are folded over the back.

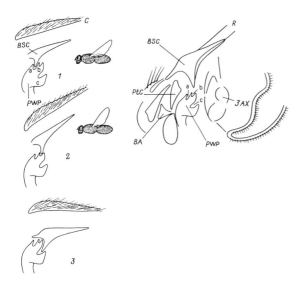

Fig. 9.18 Interaction of the tooth on the ventral side of the basisubcostale and the three points of the pleural wing process in the blow fly *Calliphora vicina*. For explanation of a, b, and c and 1, 2, and 3 see text. *PtC* is the receptor characteristic of dipteran wing articulation.

of the muscles of the third axillary sclerite this tooth 'clicks' into the posterior incision thus ensuring semi-automatic wing folding. The two anterior incisions probably correspond to standard positions of the stroke plane during flight (Fig. 9.18, 1 and 2).

The transition from roof-like to flat folding, which is apparent from certain oligoneuropteran groups was accompanied by changes in the structure of the distal region of the axillary apparatus. Characteristic features of roof-like folding include a well-developed triangular *DMP* and a high transverse *PMP*. In the Panorpidae, which fold their wings flat, *DMP* has become reduced on the forewings. *DMP* is not completely reduced on the hindwings; however, *PMP* is short, and its markedly transverse position on the wing base has been lost. This *PMP* structure is typical in Diptera (Figs 9.10 and 9.11), but small *DMP* retains its typical triangular shape. In hymenopteran wings *PMP* has become immobilized, and only a small plate is left of *DMP*, which forms a joint with the base of *PCu* (Fig. 9.12).

9.3 The increase in wingbeat frequency

Coupling of the fore- and hindwings was a prerequisite for an increase in wingbeat frequency. This increase was associated with expansion of the physiological potential of the wing muscles, which in turn affected their structure. Wing muscles of a histologically very primitive—tubular—type were replaced by the closely-packed ones typical of mayflies, some orthopterans, and most functionally two-winged insects. The highest wingbeat frequency that has been recorded in medium-sized and large insects with closely-packed wing muscles has been found in certain sphinx moths—90–95 Hz. The highest wingbeat frequencies are attained with fibrillar-type musculature. The main peculiarity of such muscles is that the frequencies of muscle contraction and of the contraction-inducing nervous impulses are different, so that such muscles are called 'asynchronous'. Asynchronous muscles are typical of the Hymenoptera, Diptera, and Coleoptera

as well as of small species from other groups (Cullen 1974).

In contrast to 'synchronous' muscles which respond to each nerve impulse with a single contraction, asynchronous muscles contract at a frequency which is timed by the muscles themselves rather than by central innervation. The frequency of muscle contraction is equal to the wingbeat frequency. Each nerve impulse accounts for as many as 5–20 wing movement cycles. Muscle contraction is initiated by the slight stretching caused by thorax deformation as a result of the contraction of antagonistic muscles—the dorsal longitudinal and dorsoventral muscles.

An increase in wingbeat frequency was associated with a decrease in wingbeat amplitude. Furthermore, a higher wingbeat frequency and hence a higher rate of up and down wing movements enabled insects to reach higher flying speeds. In addition, increased wingbeat frequency was associated with a decrease in average stroke plane angle relative to the horizontal and liberation of the stroke plane orientation relative to the longitudinal axis of the body. Parallel to the evolutionary increase in wingbeat frequency and decrease in stroke amplitude there was an increase in the angle of wing rotation around the longitudinal axis of the wing during pronation and supination. This change resulted from a decrease in stroke plane angle relative to the longitudinal axis of the body. In Diptera the angle of wing rotation around its longitudinal axis can reach 180°, especially while manoeuvring (Fig. 9.19). Moreover, mastery of high wingbeat frequencies predisposed insects to greater use of the skeleton's elasticity and resulted in improvements in the operation of the axillary apparatus in addition to the changes in wing morphology and functioning.

9.3.1 Adaptations for wing deformation

The highest wingbeat frequencies are achieved in the Diptera and Hymenoptera. This type of high frequency wing kinematics is characteristic of these insects.

Dipteran wings rotate strongly around their longitudinal axes during the strokes. The wing plane itself does not twist, as the torsion is localized at the narrow base. Already in the Nematocera the wing base is sharply narrowed and elongate, forming a 'manubrium'. The longitudinal veins in the wing base have moved closer to each other and are separated by deep folds (Fig. 9.10). The clavus, which controls wing torsion in other insects, is absent in Diptera. Only a narrow chitin strip of PCu is retained at the base of A_1 (Figs 9.10, 9.11, and 9.15), but the transverse vein $cu-pcu$ (Fig. 9.15), through which in the Nemestrinidae, Bombyliidae, and Asilidae the postcubital trachea crosses into CuP, is still clearly expressed. The cubital fold is retained at the wing base (Figs 9.10 and 9.11).

The main role in wing deformation is played by the two folds, pm and $pcua$. $pcua$ emerges immediately after Cu forks into CuA and CuP (Fig. 9.15). This fold appears as an abrupt longitudinal concavity behind CuA. In most Diptera,

the $pcua$ ridge is sclerotized, so that this fold looks like a vein, which is usually called CuP. Wootton and Ennos (1989) have recently adduced convincing evidence that the sclerotization along $pcua$ is not homologous to CuP, but is rather a pseudovein which facilitates wing deformation.

The pseudovein morphologically establishes the concave ridge in the cubital field, thereby maintaining the rigidity of the posterior part of the wing. An additional vein performs a similar function in the anterior part of the syrphid wing (Fig. 9.20a). Thus, the wings of many Diptera, especially those of the Orthorrhapha and Syrphidae, have two rigid fields, the anterior and posterior. The fields are separated from each other by a medial fold which is closely connected to the arculus, a complex hinge formed by the basal segments of CuA and M (Figs 9.20 and 9.21). Wootton and Ennos (1989) described the action of the arculus during a stroke as follows (Fig. 9.21):

The aerodynamic force, acting behind the torsional axis, tends to raise the remigium, and to impart a nose-down twist to the leading edge

Fig. 9.19 The tachinid fly *Echinomyia grossa* taking off backwards and sideways. Notice that to achieve this extraordinary manoeuvre the leading edges of the wings have twisted through almost 180°, so that the trailing edges point forwards (after Dalton 1975).

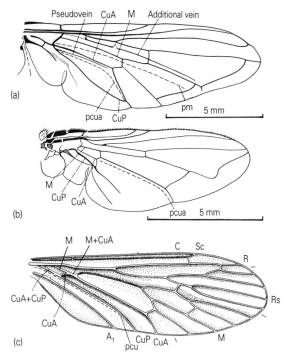

(a)

(b)

(c)

Fig. 9.20 Functional wing design in the Diptera. (a) Wing of the hover fly *Syrphus* sp. (b) Wing of the house fly *Musca domestica*. (c) Hypothetical dipteran wing ground plan (after Wootton and Ennos 1989).

spar, from which the remigial veins diverge. This automatically generates a cambered aerofoil, which is enhanced by the presence of the arculus . . . The lifting of the media, which is attached to *CuA* at the arculus, and linked to it more distally by a stiff cross-vein, tends to twist the spar formed by *CuA*, *CuP* and the intervening pseudo-vein in a tail-down sense, forcing down the trailing edge, enhancing the camber of the aerofoil, and maintaining a positive angle of attack.

The arculus mechanism acts most efficiently during the downstroke (when *M* is moved up by aerodynamic force), and causes wing pronation in the Diptera Orthorrhapha and Syrphidae, which occurs by longitudinal bending along the medial fold. In the Diptera Cyclorrhapha,

whose wings are not clearly divided into two rigid fields (Fig. 9.20b), the action of the arculus mechanism is weaker. However, discontinuities in the supporting veins (*C* and *R*), the presence of several torsional hinges at the wing base, as well as the softness and flexibility of the rear margin of the wing are all conducive to wave propagation from tip to base during wing pronation. The same torsional hinges cause wing deformation in Diptera during the upstroke. As the veins which support the wing tip come at it obliquely from the wing base, where they are hinged on to *C* and *R* by the arculus, the anteroapical tip region bends down at the bottom of the stroke. As the wing tip returns to its starting position, a wave which causes supination of the whole surface runs from the wing tip along the rear edge of the wing towards the body (see Fig. 2.17, p.47). This type of supination, a 'supinational wave', is characteristic of wings with a narrow base and elastic tip region behind which the wing is soft and flexible, thus favouring wave propagation. The velocity of wave propagation along the wing and wave amplitude depend on both wing width and flexibility.

In the Diptera which have relatively broad wings, the wave travels slowly, with a large amplitude. Maximum wave propagation velocity is attained by the relatively narrow and stiff wings of certain Diptera Orthorrhapha (see Fig. 2.10, p.42), and can be regarded as a separate

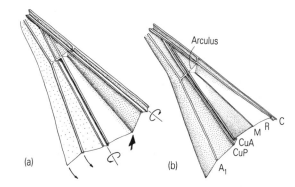

(a)   (b)

Fig. 9.21 Mode of action of the arculus (after Wootton and Ennos 1989). The thick arrow indicates the point at which the aerodynamic force acts upon (a), causing the wing plate to fold as in (b).

type of supinational deformation, 'supinational rotation'—a very quick, small-amplitude rear marginal wave.

When the wave reaches the body, the wings are completely twisted at the base, and the wave begins to propagate in the opposite direction. As a result the wing straightens and continues to move upwards while slightly supinated. The areas of the wing in which resilient deformational energy is accumulated are of particular interest. These regions act as a trigger mechanism which initiates the deformation that brings the wing back to its starting position. The anal lobe of the crane fly *Cylindrotoma distinctissima* (see Fig. 2.17, 10–16, p.47) exemplifies such regions.

In the Hymenoptera the hindwing is firmly coupled along its entire costal edge to the forewing by three groups of hook-like bristles. An oblique fold has the principal role in deformation of the wings (Fig. 9.22). At the bottom of the stroke the wing tip, loaded with the pterostigma, bends down along the oblique fold. The wing tip returns to its original state by means of an abrupt turn, which initiates a wave which runs along the wing from tip to base (see Fig. 2.8, 10–13, p.48). At the very end of the upstroke the resilient deformational forces straighten the supinated wing such that at the beginning of the downstroke those forces continue their action, thus in turn causing pronational bending of the wing tip (see Fig. 2.8, 13–16 and 1 and 2, p.48). At the beginning of the downstroke the wings appear to be fully pronated (see Fig. 2.8, 3). This sort of wing deformation during pronation and supination is typical of the lower Hymenoptera, which have relatively broad, soft wings, and especially of the ichneumon *Ophion* sp. In the Hymenoptera Aculeata, the wings are narrower and more rigid; the supinational wave during the upstroke runs faster along these wings, and during the downstroke the coupled wing pair on one side bends along the coupling line (Fig. 9.22a). Hence wing deformation during the upstroke is similar in the Diptera and Hymenoptera. The deformation looks like a supinational wave running along the wings from the tip to the body. During the downstroke, wing pronation in the lower Hymenoptera,

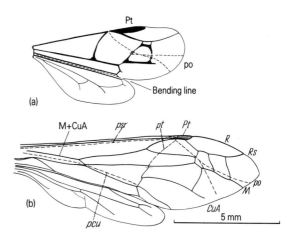

Fig. 9.22 Functional wing design in the Hymenoptera. (a) The organization of a coupled wing pair which functions as a flapping surface (modified from Bocharova-Messner 1982). (b) Wing pair of the common wasp *Vespula germanica*.

Diptera Nematocera, and Diptera Brachycera Cyclorrhapha (except Syrphidae) occurs as a wave which propagates from the wing tip to the body. The pronation process is quite different in the Hymenoptera Aculeata, Diptera Brachycera Orthorrhapha, and Syrphidae. In these insects the wing or the wing pair bend longitudinally during pronation, forming a convex profile.

9.3.2 Flight aerodynamics

The pattern of vortices formed during the flight of insects with a high wingbeat frequency is strikingly different from that of insects with lower wingbeat frequencies (see Chapter 3). During tethered flight of the crane fly *Tipula paludosa*, two (instead of one) chains of vortex rings were observed. Detailed investigation of this phenomenon (Brodsky 1986) uncovered significant details about the formation and shedding of the vortex rings from the wings. The flapping cycle in the crane fly begins with wing pronation; the wing tip bends downwards, causing intensive vortex formation. As the wing goes down, the tip vortex increases quickly and there is an immediate increase in airflow in the proximal part of the wing. This current rolls into a basal vortex which rotates towards the tip vortex; the two vortices then accelerate a narrow air current which flows backwards and crosses the wing at its broadest part. By the end of the downstroke, the small diameter vortex ring is closed through tip and basal vortices on the dorsal surface of the wing (Fig. 9.23a).

Shedding of rings from the wings and their final closing occur at the bottom of the stroke, and are accompanied by a strong air current moving symmetrically from the upper and rear sides distally and downwards (Fig. 9.23a). As the wings begin the upstroke, the rings formed during the downstroke move downwards. By the end of the upstroke the vortex rings have moved backwards by a distance approximately equal to the length of the abdomen. As a result, an aerodynamic wake consisting of two chains of

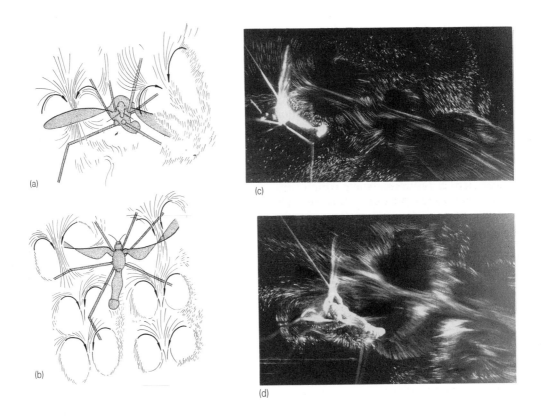

Fig. 9.23 Airflow around the crane fly *Tipula paludosa* in tethered flight. Stroke parameters: $n = 58$ Hz, $A = 100°$, $b = 50°$, $B = 30°$. The tethered crane fly is trying to turn to the left. (a) Front view, using a vertical light plane passing through the wing bases. Note the vortex ring closing on the wing of one side, at the bottom of the stroke. (b) View from above, using a horizontal light plane passing through the wing bases. Note the two chains of small uncoupled vortex rings. (c and d) Photographs of the crane fly, using a vertical light plane passing through the basal third of the left wing, side view, the head to the left. Note (c) the left-hand chain of small uncoupled vortex rings and (d) the airflow passing straight through their centres.

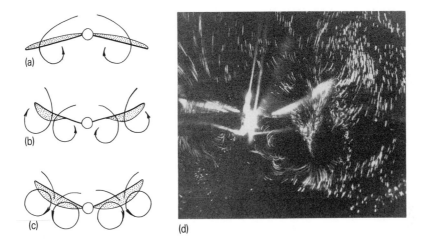

Fig. 9.24 (a–c) Diagrammatic representation and (d) a photograph of the air currents around the crane fly showing (a–c) successive stages in formation and (d) shedding of vortex rings during the upstroke. Front view, with a vertical light plane passing through the wing bases.

small uncoupled vortex rings is left behind the flying insect (Fig. 9.23b). A cross-section through these chains using a vertical plane of light which passes through the basal third of the wing renders two distinct and a third dissipating vortex ring visible on the left side of the body (Fig. 9.23c), and also a current which passes straight through their centres (Fig. 9.23d).

At the lowest point in the stroke, the air current which caused shedding of the rings forms a stopping vortex basally, which rotates in the same direction as the corresponding vortex of insects with a low wingbeat frequency—from the wing base distally and downwards (Fig. 9.24a). As the upward movement of the wing begins, the vortex formation occurs in the area of the wing tip due to the tip-to-base wave, and consequently each stopping vortex contributes to the missing part of the ring (Fig. 9.24b and c). Upward movement of the wing is accompanied by a deformational wave which runs from base to tip, causing ring shedding from the wing (Fig. 9.24d); the rings dissipate by the end of the upstroke, leaving a slight air vorticity beneath the insect's body (Fig. 9.23d).

This unusual vortex wake shape in the flight of the crane fly can be explained by wing kinematics and deformational features. The first condition for vortex ring formation on the wing on one side of the body is the domination of the tip vortex in the airflow around the wing. At the beginning of the up- and downstrokes, the wing tip is at first bent, thus affecting the flow and causing vortex formation.

In ichneumons, Diptera Nematocera, and some Diptera Brachycera, pronation is accomplished by a wave: first of all the wing tip bends down, and then this bend propagates as a single wave towards the body, causing pronation of the whole wing plane. With this sort of pronation the wing tip is first to begin influencing the flow.

In 1973 Weis-Fogh proposed a mechanism which might underlie lift formation—a 'flip-mechanism', based on the ability of the wings of Syrphidae to bend longitudinally along the medial fold. The existence of such a mechanism could not be confirmed subsequently; however, the idea of longitudinal wing bending during pronation seems to be important for explaining the very early appearance of a vortex at the tip as the fold, along which bending occurs joins with the anterior wing margin to form a triangle which widens towards the tip (Fig. 9.20a). Therefore, even if pronational bending occurs in its usual manner, evenly along the anterior margin, the broader bending part of the wing tip will cause greater air turbulence than the proximal wing region.

In addition, in insects with high wingbeat frequencies there is no wing clap at the upper point of the stroke. A wing clap above the tergum facilitates the formation of a common ring behind the wings of the right and left sides of the body, since as the wings move apart after the clap the air rushes into a gap between the parted leading edges, resulting in vortex formation on the dorsal surface of each wing.

The tip vortex of a flapping wing differs from that of a supporting gliding surface which functions under steady conditions. In the latter case a tip vortex is formed as a result of a current flowing from the high pressure region on the lower surface of the wing to the low pressure region on the upper surface. In consequence, the main plane of tip vortex rotation appears to lie approximately perpendicular to the supporting plane. Although the tip vortex of the supporting plane interferes with the aerodynamic quality of the wing, a vortex formed at the tip of a flapping wing is part of the vortex ring and is therefore a necessary condition for the creation of beneficial aerodynamic forces. The orientation of the tip vortex rotation plane depends on the stroke plane angle relative to the longitudinal axis of the body. With a sheer stroke plane, the air currents run from the lower surface to the upper one. As the stroke plane angle decreased, the plane of tip vortex rotation approaches closer and closer to that of the wing itself. This evolutionary trend becomes more distinct with increased wingbeat frequency, and creates favorable conditions for tip vortex domination in the air currents around the wing.

Thus, closing of vortex rings on the wing (or the coupled wing pair) on one side of the body is favoured by:

(1) high wingbeat frequency;
(2) low wingbeat amplitude;
(3) a small stroke plane angle relative to the longitudinal axis of the body;
(4) the absence of the wing clap above the tergum at the upper point of the stroke;
(5) an unusual method of wing deformation at the upper point of the stroke, either as a pronational wave, as in the crane fly wing, or as pronational bending, as in the hover fly wing.

However, it would be a mistake to believe that the flight of all insects with a high wingbeat frequency invariably results in the formation of two ring chains. The description above of the mechanism by which vortex rings close on the wing on one side of the body is intended more to indicate the theoretical possibility that this process may occur. Insects with high wingbeat frequencies can actually make use of a number of vortex ring formation methods. Thus, a crane fly manoeuvring among plants uses a low amplitude wing stroke, whereas, when hovering or flying upwards the stroke amplitude increases. When the wings are moving with a large amplitude, the stopping vortices of the right and left wings unite to form a common vortex ring.

For the vortex ring to be able to close on the wing, a second condition is needed: that the maximum wing width (the chord length), i.e. the place where the airflow crosses the wing, should be as far as possible from the body. This condition is associated with the need to prevent the destruction of the basal vortex by interference from the body. The relationship between the wingbeat frequency necessary for rings to be formed by the wing and wing morphology, particularly when there is significant wing deformation during the stroke, is an important problem that needs special investigation. At present, it can be noted that most insects with a high wingbeat frequency have narrow elongate wings whose maximum chord length falls almost at the middle of the wing. Where the wings are broad, various devices have developed which protect the proximal part of the ring from interaction with the body. One example of such a device is the alula in the jugal wing region in some Diptera and Hymenoptera. In the Diptera Cyclorrhapha a similar function seems to be performed by a soft elastic anal lobe. It is probable that the alar squamae of higher flies to which muscles are attached play an aerodynamic role, protecting the ring from becoming dispersed at certain points in the stroke.

The mastery of a new method of vortex ring closing was an important qualitative leap in the development of insect flight. First of all, the independent operation of the right and left wings became possible, which led to a striking increase in flight manoeuvrability. Nachtigall (1979) showed that in the higher Diptera the wings on the two sides can work in phase, phase-shifted, out of phase, and even with one wing temporarily switched off. The advantage of this is obvious. For instance, a hover fly has only to change the stroke plane of one of its wings slightly for a preponderance of forces to appear on one side of the body, so that the insect performs a sudden sideways jerk. In insects with low wingbeat frequencies only dragonflies can carry out such a manoeuvre.

Another advantage which results from the new mechanism for the production of aerodynamic forces is the independence of the strength of the force produced from the stroke amplitude value. Indeed, the decrease in the stroke amplitude and consequently of area swept by the wing actually led to a reduction in flight efficiency. However, in insects with high wingbeat frequencies these losses are fully compensated for both by increasing the flow velocity in the wake and by involving the elastic forces of the skeleton in wing operation. These forces can be brought to bear twice, at the lower and the upper points of the stroke. Thus it is clear that the decrease in wingbeat amplitude produces favourable conditions for the employment of the elastic forces of the skeleton, even leading to the appearance of the phenomenon of resonance in the wing-body system. This explains the unexpected dependence of the lift produced by the fly *Calliphora* on its wingbeat frequency, and the independence from its stroke amplitude (Nachtigall and Roth 1983).

9.3.3 Flight behaviour of insects with high wingbeat frequencies

At equal body sizes, flight velocity increases as wingbeat frequency increases. High velocity and manoeuvrability are distinguishing features of insects with high wingbeat frequencies. The flight of Diptera Cyclorrhapha has the highest velocity and manoeuvrability. Frequent banking turns are accompanied by rapid turning of the longitudinal axis of the body into the direction of flight. However, the higher flies are capable of flying sideways, with their legs pointing upwards, making sudden stops, and prolonged hovering. For instance, the house fly can stop instantly in mid-flight, hover, turn itself around its longitudinal body axis, fly with its legs up, loop the loop, turn a somersault, and sit down on the ceiling, all in a fraction of a second. Similar flight behaviour occurs in certain Orthorrhapha, such as the Stratiomyidae.

The Diptera which fly among herbaceous plants such as the Empididae, Asilidae, Rhagionidae, and Syrphidae have highly manoeuvrable flight with frequent directional changes. All their manoeuvres are performed at high speed. Frequent hovering and abrupt jerks to the side are typical of these insects. Either hunting or searching for a mating partner in the undergrowth requires a high degree of manoeuvrability, but without significant displacement. This is achieved by frequent use of hovering. Some hover flies have developed a second, more advanced form of hovering.

As Collett and Land (1975) have shown by filming the movements of hover fly *Syritta* in the horizontal plane, these animals are able to fly forward, backward, and sideways independently of angular movements. In fact, they seem to be able to adjust any ratio of sidewise to forward motion. The manoeuvrability of house flies is more restricted due to higher inertia. House flies turn by banking (Wagner 1985): when entering a curve the body is rolled to lower the side pointing into the centre of the curve—as the pilot of an aeroplane would do when flying a curve. This movement is accompanied by a rotation about both the vertical and transverse axes. The angular velocity about these two axes rises and falls steeply, but does not change in sign, whereas the rolling movement is characterized by a change of sign of the angular velocity and lasts longer than the movement about the other axes. Coming out of the curve, roll angle and pitch (body) angle return to their original values. Thus only a change in the orientation of the long axis results. For a flying 'machine' that cannot produce an active sidewise thrust, it is especially advantageous to be able to control roll and yaw independently of each other. By doing so, manoeuvrability is increased.

Many Hymenoptera and Diptera, such as Tabanidae and Nemestrinidae, which inhabit exposed areas are characterized by swift rectilinear flight. These insects also hover when necessary, but less frequently than in the Diptera considered above. When searching for large, distant objects they can cover long distances rapidly. Changes in flight direction are accompanied by a rapid turning of the body, without stopping.

Long-distance active migrations are not usual in the Diptera and Hymenoptera: swarming behaviour is far more typical. Swarms are formed

by numerous dipteran species: mosquitoes, dance flies, biting midges, horse flies, gall midges, crane flies, fungus gnats, and march flies; yet the most striking swarming occurs in midges. The mass swarming of these insects is exceeded only by mayflies. In swarms of some dance flies (Empididae) sometimes one of the males leaves the swarm, bursts into a neighbouring mosquito swarm (both groups of insects often swarm above the same marker, e.g. a stump or a rubbish-heap), seizes a mosquito, and holding it tight with its forelegs, returns to its own swarm. Then the male dance fly is ready to meet a female. The male, with its 'nuptial gift', hovers at different points in the swarm, keenly responsive to anything above it. At times the male may move upwards but, after discovering that a male is hovering above, returns to its original position. However, if the insect overhead turns out to be a female, the male delivers the nuptial gift and begins to copulate with the female in the air. The mating couple quickly descend on an inclined trajectory and land on the substratum.

9.4 The decrease in wingbeat frequency

Although the principal evolutionary trend in the flight apparatus of Oligoneoptera was towards an increase in the wingbeat frequency, a decrease in wingbeat frequency has occurred in some Neuroptera and Lepidoptera. This direction of specialization of the wing apparatus seems to have been the most promising in the Lepidoptera, having resulted in a wide variety of forms.

Kozlov, Ivanov, and Grodnitsky (1986) have followed the changes in wingbeat frequency and wing shape in various evolutionary lines of the Lepidoptera (Fig. 9.25). The wings of the most primitive Lepidoptera, such as the Micropterigidae or Eriocraniidae, are elongate, oval, and symmetrical relative to the basoapical axis (Fig. 9.25a). In the Hepialidae the tornal angle has emerged and the wing plane has lost the symmetry originally typical of these insects, and approaches the triangular wing common for most of insects. The unification of the fore-

and hindwings into an integral flapping surface served as a prerequisite for wide evolutionary and adaptive transformations of the wing apparatus in Lepidoptera. Removal of the functional load from the costal edge of the hindwing led to changes in both its shape and venation.

One of the directions of wing transformation in the Lepidoptera, narrowing of the wing plane and the development of a long fringe, is not limited to a small group of families, although it occurs largely in primitive moths whose wingspan does not usually exceed 20 mm (Fig. 9.25b–d). The wings of narrow-winged moths are completely coupled and work as an integral flapping surface; this can be seen from the deformation waves which pass smoothly from the forewing to the hindwing. The curious structure of the lobed wings of the Pterophoridae (Fig. 9.25g and h), and Alucitidae (Fig. 9.25f), which suggests that the flight of these moths is quite out of the ordinary, drew special attention to these Lepidoptera from numerous investigators (e.g. Norberg 1972*a*, and Ellington 1984). However, contrary to expectations, no significant difference could be found between the flight of these moths and that of moths with smooth-edged wings.

Another direction in which the wing apparatus changed, narrowing of the fore- and hindwings without formation of a long fringe (Fig. 9.25j), is clearly noticeable within the limits of the natural group of superfamilies comprising the Cossoidea, Zygaenoidea, and Sesioidea.

The third way in which the wing apparatus changed can be divided into two directions of wing specialization: either that of the forewing narrowing and hindwing reduction, or that of the forewing broadening and hindwing enlarging and becoming rounder. The line leading to hindwing reduction is most noticeable in the Noctuidae (Fig. 9.25i), Notodontidae, Sphingidae (Fig. 9.25k), Syntomidae, and Thyrididae. In the most fast-flying moths of this group—the sphinx moths and some owlet moths (Cuculliinae)—not only has the wing apparatus been modified, but the whole body has become streamlined. Due to the preferential development of the mesothorax, the pterothorax is strengthened; the metathoracic muscles serve primarily to move the legs but not the wings.

Wing broadening, a second mode of wing specialization, is apparent in the Papilionoidea, Saturnioidea, Geometroidea, and some other groups. The areas of the fore- and hindwings are virtually equal (Fig. 9.25 l, m, and n), and the body size is very small compared with the wings. Costalization of the forewings is very pronounced. In representatives of many groups (Hesperoidea, Papilionoidea, etc.) the frenulum is reduced and wing coupling is achieved by the overlap of the anterior margin of the hindwing by the rear margin of the forewing, the scale cover being strongly modified in the contact zone. The principal characteristic of the flight of broad-winged butterflies is their low wingbeat frequencies, which range from 30 Hz in the skipper *Thymelicus lineola* to 8 Hz in some Saturniidae.

Skippers and butterflies are united in the suborder Rhopalocera. In these insects the medial stem has disappeared from both fore- and hindwings, which has resulted in the appearance of a large medial discoidal cell. Most species are brightly coloured. The imagos are active in daylight and prefer sunny weather; their broad wings collect the warmth of the sun and serve as efficient thermoregulators. At rest, the Rhopalocera keep their wings erected over the tergum rather than folding them flat, as do nocturnal moths. A large body size and broad wings has predisposed these insects to the mastery of gliding flight.

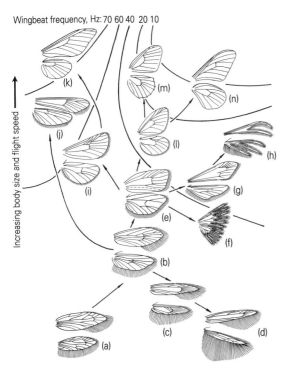

Fig. 9.25 Main wing shape types and their possible evolution in the Lepidoptera (after Kozlov, Ivanov, and Grodnitsky 1986). (a) Micropterigidae, (b) Adelidae, (c) Tineidae, (d) Tischeriidae, (e) Tortricidae, (f) Alucitidae, (g) and (h) Pterophoridae, (i) Noctuidae, (j) Sesiidae, (k) Sphingidae, (l) Geometridae, (m) Saturniidae, (n) Papilionidae.

9.4.1 Adaptations of butterflies for gliding flight

The adaptations for gliding flight have involved the structure of the wings and of the axillary apparatus. We shall first consider changes in the structure of the wings and their scale cover.

Gliding flight occurs in numerous lepidopteran species, especially in those with a low wing loading (see Chapter 4). In certain species of Lepidoptera gliding flight accounts for 40–50 per cent of their overall flying time. During gliding flight butterflies hold their wings in a particular position, as follows: the forewings are pulled backwards and slightly lifted, while the hindwings are moved close together, with their rear edges under the abdomen (Fig. 9.26). This position probably allows the butterfly to increase its gliding flight stability when pitching by shifting the aerodynamic pressure centre backwards. Several significantly differing types of lepidopteran outlines can be distinguished.

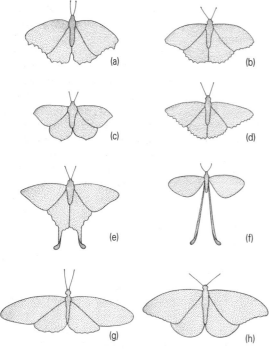

Fig. 9.26 Wing and body outlines of neuropteran *Nemoptera sinuata* (f) and various butterflies when gliding: (a) *Nymphalis antiopa*, (b) *Argynnis paphia*, (c) *Gonepteryx rhamni*, (d) *Vanessa cardui*, (e) *Papilio podalirius*, (g) *Heliconius charithonius*, (h) *Danaus plexippus*.

First of all, the butterflies have different wing aspect ratios. Butterflies with a maximum aspect ratio are exemplified by the '*Heliconius*' type (Fig. 9.26g), whereas those with a minimum aspect ratio are represented by the '*Papilio*' type (Fig. 9.26e). The hindwings of butterflies with a minimum aspect ratio usually have additional processes—so called 'spurs' or 'tails'.

The classical glider wing shape (with a high aspect ratio) is most closely approached by the *Heliconius* wings; however, even in this butterfly the aspect ratio, based on the wing pair on one side, does not exceed $\lambda = 2$, whereas in a glider it is about 7 to 12. The wings of most other butterflies which are capable of gliding have even lower aspect ratios. The aerodynamics of a low aspect ratio plane were studied by Golubev (1957). This author demonstrated that:

(1) at low aspect ratios, there was a sharp increase in the critical angle of attack (this increase being from 10–12° with an aspect ratio $\lambda = 5$, to 35° with $\lambda = 1$);

(2) a low aspect ratio wing was strongly influenced by tip vortices: these tip vortices twist and accelerate the flow on the upper surface (Fig. 9.27a);

(3) the influence of the airflow over the lateral edges became significant only with low λ values and high angles of attack;

(4) a low aspect ratio wing incurred a comparatively strong induced drag.

Martin and Carpenter (1977), using stiff models and dry collection specimens of *Papilio machaon* and *Pieris brassicae*, observed flow lines (Fig. 9.27b) similar to those obtained with a low aspect ratio plane (Fig. 9.27a). The tip vortices appeared to affect the airflow around a butterfly such that a portion of the flow deviated from its original direction and rolled up into two vortex braids, whose direction of rotation was opposite to that of the tip vortices (Fig. 9.27b(i)). Viewed from the side (Fig. 9.27b(ii)), the tip vortices maintain an orientation close to that of the oncoming air, and the vortex braids which flow down from the inner border of the wings bend downwards.

The diagrams of flow lines obtained by Golubev and by Martin and Carpenter may be used

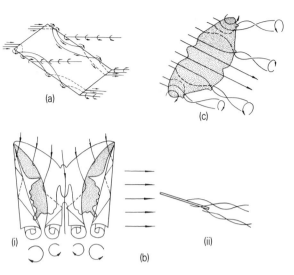

Fig. 9.27 Airflow around a low aspect ratio plane and around butterflies. In all cases the airflow down the wings forms two pairs of braids, in each of which the vortex braids are twisted in opposite directions. (a) Low aspect ratio plane (after Golubev 1957). (b) Model of swallowtail butterfly with extended wings: (i) plan view and (ii) lateral view, showing the left-hand-side pair of vortex braids, with the tip one above (after Martin and Carpenter 1977). (c) Camberwell beauty butterfly: same view as in (a).

when considering the characteristics of the airflow round a 'Papilio' type gliding butterfly. However, it should be remembered that *Papilio*-type butterflies have a somewhat higher aspect ratio when they draw their wings backwards than do the butterflies studied by Martin and Carpenter which had their wings spread out. Bearing this in mind, we shall now compare the airflow around a low aspect ratio plane with that of a gliding butterfly. As the air flows around the low aspect ratio plane, the tip vortices first affect the direction of flow on the lower surface (Fig. 9.27a). As a result, the airflow on the lower surfaces of the butterfly's wings veers towards the tip in the distal wing regions, and towards the body in the proximal ones (Fig. 9.27c). Each of the tip vortices induces the formation of a complementary vortex, so that the flow is divided into tip and root sections

(Fig. 9.27c). In the proximal zone of the upper surface, the air currents cross the wing plane almost without any change in direction; only near the border do they veer off slightly distally and flow down along the marginal grooves (Fig. 9.27c). As the proximal part of the airflow along the lower wing surface moves closer to the body than the corresponding airflow on the upper surface, both airflows become twisted into a vortex braid after leaving the wing. In the distal region of the wing, the air flows along grooves which are oriented the same way as the marginal veins. Hence the airflows around a gliding butterfly are divided to form two pairs of vortex braids: a pair of tip braids and a pair of braids on the hindwings' terminal margin.

The most complete airflow division is to be expected in butterflies with the minimum wing aspect ratio. These (*Papilio*-type) butterflies are provided with long tails on their hindwings. The orientation of these tails during gliding flight (that is, pointing straight backwards (Fig. 9.26e)), and the direction in which they are twisted, correspond with the orientation and direction of twisting of the inner pair of vortex braids which flow off the low aspect ratio plane (Fig. 9.27a) and the wings of the *Machaon* model (Fig. 9.27b). The features noted above may indicate that each tail serves as a kind of a 'catcher' of vortex braids as they flow off the proximal half of the upper and lower wing surfaces. The extremely long tails of the lacewing *Nemoptera sinuata*, as well as those of other gliding neuropterans, may serve as a tail unit that increases the insect's ability to maintain its course during gliding.

From the orientation of the vortex braids behind the gliding butterfly (Fig. 9.27b(ii)) we can assume that a considerable component of total drag is the drag caused by the tip vortices. The tip vortices are more pronounced with short, broad wings than with narrow, long wings of an equal area. Consequently the cross-sectional area of the tip vortex is less in butterflies that have elongate, sharp-tipped wings. Furthermore, by lifting the wing tips in such a way that the angle between the surfaces of the right and left wings is less than 180°, the gliding butterfly can reduce the tip vortex intensity and thus the induced drag, simultaneously increasing its stability in the rolling plane (Zalessky 1955). Another important source of drag is the profile drag, which is proportional to the cross-sectional area of the inner pair of vortex braids (Fig. 9.27c). Martin and Carpenter (1977) have shown that when the angle of attack is greater than a certain value, the inner pair of braids is replaced by a wide zone of turbulence behind the body, which is accompanied by a substantial increase in drag. This is not, however, the case when the wing tails function as catchers so that their track still consists of two vortex braids twisted in opposite directions; this allows the tail-bearing butterflies to use greater angles of attack during gliding without significant increase in drag.

However, nowhere near all the butterflies which are capable of gliding have tails on their hindwings. In particular, the tails are absent from many species with a low wing aspect ratio. This requires attention to be paid to the difference in organization between the upper and lower wing surfaces. The proximal area of

the upper wing surface is organized in the same way in most butterflies: the sloping longitudinal ridges formed by the veins project slightly beyond the membrane surface. The entire surface is covered with dense even rows of scales which overlap like tiles, the scales covering the veins obliquely (with their ends towards the wing tip) and thus smoothing out the ridges. In general, the upper outline of the wing profile is wavy in the proximal zone. Nearer to the insect's body the wing surface has a dense covering of long, soft hairs (Fig. 9.28a(i)) which change orientation even in slight air currents. In the posterior part of the proximal zone the hairs wrap round the abdomen, covering it from the sides. The distal zone of the wing surface forms a system of grooves running along the folds, oriented perpendicular to the terminal margin. The scale cover is even; the position and orientation of the scales in the perimarginal zone obey the general rules of wing outgrowth orientation, and depend on the orientation of

Fig. 9.28 The different organization of (i) the upper and (ii) the lower wing surfaces of two butterfly species. (a) Camberwell beauty, *Nymphalis antiopa*, (b) brimstone, *Gonepteryx rhamni*.

ridges and grooves (see Fig. 1.20, p.30).

The lower surface is organized differently. Deep furrows form between the strongly projecting ridges of the longitudinal veins which mostly lie across the oncoming airflow. Long, coarse hairs arise from the bottom of these furrows, sometimes with capitate bulges on the tips. These hairs seem to prevent stalling as a result of the projection of the sharp veins. Trial wing models of an anisopterous dragonfly (Newman *et al.* 1977) and some Lepidoptera (Buckholz 1986) in a wind-tunnel have shown that stalling did occur on the ridges of projecting veins, which led to the formation of small stable vortices in the depressions just behind the ridges (i.e. downstream). The hairs in these fissures and furrows prevent this stalling and the area in which there are projecting hairs is evenly rough. Taken as a whole, the lower outline of the proximal region of the wing can still be described as furrowed. At the back of this region there are numerous hairs, which are much shorter and coarser than those on the upper surface, and which spread more distally along the veins from the wing base. In the distal part of the lower wing surface there are well developed grooves which are shorter, denser, and deeper than those on the upper surface, as the folds among the marginal veins on the lower surface project over the membrane like bolsters.

The greatest difference in organization between the upper and lower wing surfaces occurs in the Papilionidae: the veins on the lower surface are not scaled, and project sharply over the membrane surface, while the hairs on the membrane itself have their tips slightly inclined towards the wing tip. In *Parnassius apollo, Papilio machaon*, and *Parnassius mnemosinae* the scale cover is thinned out and individual scales are spear-shaped. In other butterfly families, the lower wing surface may be roughened in various ways. Thus, in *Pieris brassicae* the strongly projecting veins of the lower wing surface have no scale cover, whereas in *Gonepteryx rhamni* these veins are covered with long scales which protrude noticeably against the background of the general scale cover (Fig. 9.28b(ii)). In *Nymphalis antiopa* the lower wing surface has rows of tough curved bristles (Fig. 9.28a(ii)) which increase the roughness of the lower wing surface. In the pierid butterfly *Phoebis sennae*, the wings are covered with scales of a similar shape and size; however, along the anterior margin of the discoidal cell on the lower side these scales do not overlap like roof tiles, but stand upright instead.

In order to analyse the impact of the difference in roughness between the upper and lower wing surfaces on their aerodynamic characteristics, Brodsky and Vorobjov (1990) studied changes in lift and drag coefficients in a wing model with symmetrical profile and with different types and distributions of roughness on the lower surface (Fig. 9.29). Airflow over the rough surface gives rise to an aerodynamic cross-wind force directed away from the rough surface, this force being greater the rougher the surface (Fig. 9.30). The increased roughness of the lower surface of a wing fixed at a zero angle of attack raises the lift to the same value as for same wing, with both surfaces smooth, at an angle of attack of 3–4°. This effect is manifest from a zero angle of attack up to two-thirds of its critical value. The lift increase is mainly influenced by the generation of turbulence, caused by the roughness, and by the distribution of the regions of greater roughness on the wing. The area of roughness which is located from 0.05–0.1 of the chord length from the leading edge of the wing to 0.35–0.45 of the chord length is the most important for the creation of this additional lift. This is where the areas with distinct hairs and scales are situated on the wing of *Parnassius apollo* (Brodsky and Vorobjov 1990). The increase in lift is also influenced by the viscosity of the medium: the higher the relative viscosity (or the lower the Reynolds number), the greater the increase in lift for the same roughness characteristics. It must be particularly emphasized that in the Reynolds number range at which the wing model was tested, all the elements providing roughness were immersed in the boundary layer, as could be shown by calculation.

If one considers the characteristics of the wing surface organization of butterflies capable of gliding a number of general adaptations may be identified, as follows. The butterflies with the highest aspect ratio (λ = 2), the Heliconiidae and certain Papilionidae, glide at relatively small angles of attack at which the elements of macrorelief projecting from the lower surface exert a relatively small influence on the flow. These elements may nevertheless cause some increase in lift. Butterflies with lower wing

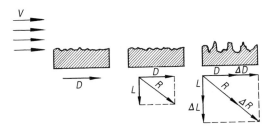

Fig. 9.30 The mechanism by which increasing roughness generates a greater lift force.

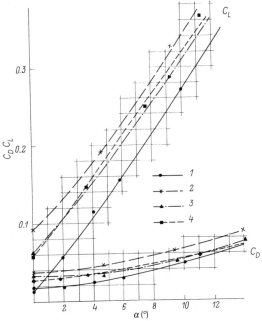

Fig. 9.29 Influence of the roughness of the lower wing surface on aerodynamic coefficients at different angles of attack. The experiments were all carried out in a wind-tunnel with a wing model of symmetrical profile and with a smooth upper surface. The Reynolds number, based on wing chord and gliding speed, ranged from 1×10^4 to 3×10^4. It has been calculated that at these values, all the roughnesses on the lower surface would be immersed in a boundary layer. The distributions of roughness on the wing were as follows: 1, both upper and lower wing surfaces smooth; 2, roughness distributed evenly over the entire surface; 3, roughness at 0.1–0.6 of a chord length from the leading edge; 4, roughness distributed evenly over the entire surface, and there were also more marked protuberances at 0.1–0.6 of a chord length from the leading edge.

aspect ratios glide at larger angles of attack. In these butterflies the difference in roughness between the upper and lower wing surfaces is more pronounced than in the Heliconiidae, and the roughness is patchy, consisting of scales and hairs that fill the furrows on the lower surface of the wing. At the minimum wing aspect ratio ($\lambda = 1$), there are two different ways to increase lift during gliding. The first one is to have long tails on the wings which provide the most efficient airflows when gliding at large angles of attack (about 35°). It is important to emphasize that in this case the difference between the roughnesses of the upper and lower wing surfaces may be more or less pronounced, and is not actually necessary, as the angle of attack is too large. The second way to increase lift is to maximize the difference in roughness between the two wing surfaces. The strongest effect of increased lower wing surface roughness on lift occurs at angles of attack of not more than two-thirds of the critical value, which is still quite high in butterflies which have a low wing aspect ratio. It is significant that butterflies of this group glide at smaller angles of attack than butterflies whose wings have long tails. The first adaptation for gliding flight, long tails, are used by some Papilionidae, in particular *Papilio podalirius*, as well as by the lacewing *Nemoptera sinuata* (see Fig. 9.26, p.176). The second adaptation occurs in *Parnassius apollo* and *Parnassius mnemosinae*. *Papilio machaon* seems to follow both directions of specialization.

A description of butterflies' adaptations for gliding flight would be incomplete without considering the changes in the axillary apparatus.

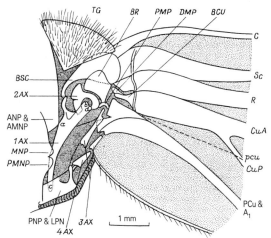

Fig. 9.31 Base of the forewing of the zebra, *Heliconius charithonius*.

All gliding butterflies have a broadened wing base, accompanied by hypertrophy of the first axillary sclerite (Fig. 9.31). The relative enlargement of the first axillary sclerite is achieved by elongation of its rear angle so that it makes contact with the posterior notal wing process. The border of the first axillary sclerite from angle *b* to angle *d* is narrowed, points upwards, and makes contact with the second axillary sclerite. In addition, the bases of the supporting vein stems are fused and strengthened by the transverse ridges. Adaptation for gliding flight exerted a negative influence on wing folding; only a pair of small sclerites is left of the third axillary sclerite at the base of *PCu* and A_1, which are fused (Fig. 9.31). The proximal medial plate has become reduced.

9.4.2 The flapping flight of broad-winged butterflies

The flapping flight of broad-winged butterflies was studied in the peacock butterfly, *Inachis io*, flying in a wind-tunnel (Brodsky 1991).

Nine wild-caught specimens were studied. Filming began after the butterfly assumed the flight posture, in which the wings flap regularly, the legs are folded and pressed against the body, and the antennae are held in the flight position. The maximum flight duration was about 1 h, and the air speeds were 0.6–1.0 m s^{-1}. All the specimens tested flew similarly, and the wing movement parameters were similar except for the wingbeat amplitude which varied from 120 to180° during the experiments. At 180°, the wings clapped together at the top and bottom of the stroke. Single individuals could perform wingbeats of varying amplitude. The most frequent stroke amplitudes were 120–130° and 160–180°, which are characteristic of feeding and migratory flights respectively. The experimental flights were therefore divided into two groups based on wingbeat amplitude. The most comprehensive data on vortex formation pattern were obtained for 'feeding' flights (120–130° amplitude), but we also discuss data from 'migratory' flights (160–180° amplitude).

9.4.2.1 A feeding flight. The cinefilm selected for analysis is of a sequence of tethered flight that can be regarded as a typical feeding flight. One complete stroke cycle lasts about 0.06 s, and can be divided into four phases of wing motion: pronation (Fig. 9.32, 1–5), the downstroke (Fig. 9.32, 5–8), supination (Fig. 9.32, 9 and 10), and the upstroke (Fig. 9.32, 10–12). At the beginning of the cycle, when the wings are raised to their maximum height (Fig. 9.32, 1), the dorsal surfaces clap together and the wings begin to pronate. The pronation phase takes up about 40 per cent of the cycle period, and is indicated by rapid backward movement of the vortex ring (Fig. 9.32, 1–5). During the subsequent downstroke, which lasts for 27 per cent of the cycle period, a starting vortex forms downstream of the wings (Fig. 9.32, 5–8). During this phase of the stroke the wings are lowered rapidly and wing position changes abruptly from one frame to another. At the bottom of the stroke the wings supinate quickly so as to move upwards. As soon as supination occurs, a stopping vortex forms below the trailing edges of the hindwings (Fig. 9.32, 9 and 10). The supination phase lasts for 10 per cent of the cycle period. The stopping vortex gradually increases in strength during the upstroke (Fig. 9.32, 10–12) which lasts for 23 per cent of the cycle period and is somewhat faster than the downstroke. The stroke phases do not coincide exactly with the frames as one frame can include the end of one phase and the beginning of the next one.

As the supinated wings approach their upper extreme, their leading edges are brought closer together although they may not collide. The wing surfaces are then pronated: first, the anterior edges of the wings are bent outwards, and the bend then passes posteriorly over the wings. As this proceeds, the leading edges diverge and the trailing edges are brought closer together. This type of wing surface deformation during pronation is typical of broad-winged butterflies and resembles Ellington's 'flat peel model' (Ellington 1984): it differs mainly in that the wing bending is sharper. The butterfly's forewing

bends along the medial fold line (Fig. 9.32, 2), but some additional lines (Fig. 9.32, 1), as well as the rear border of the forewing (Fig. 9.32, 3), are also involved. The gap between the right and left wings closes at the moment when the leading edges of the wings collide. Wing supination resembles pronation, except that the entire movement takes place in the opposite direction. The bending of the leading edge upwards passes from the anterior to the posterior border of the wing.

The stroke amplitude was 120–130° which, as mentioned above, is similar to that used in feeding flights. The angle between the stroke plane and the body axis was about 90°, as in free flight. The wingbeat frequency was 15–20 Hz. The body angle relative to the horizontal changed

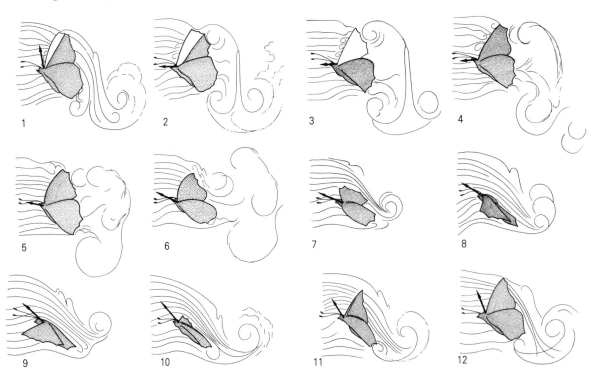

Fig. 9.32 Tethered flight of the peacock butterfly, *Inachis io* in a wind-tunnel: consecutive film tracings from a single wingbeat. The dynamics of the resultant force vector which is presumed to be acting on the butterfly is shown with bold arrow at twelve successive points in the wingbeat during 'feeding' flight. 1–5, pronation; 5–8, the downstroke; 9 and 10, supination; 10–12, the upstroke. The time between frames is 5 ms.

considerably during the cycle, and reached a maximum towards the end of the upstroke.

In Fig. 9.32, a complete stroke cycle is covered in the 12 frames. This series begins with wing pronation (Fig. 9.32, 1), as described above. The pronated part of the forewing is illuminated. At the beginning of pronation, a stable vortex, the 'leading-edge vortex', is formed along the leading edge (Fig. 9.32, 1). The air passing through the gap between the pronating wings moves sharply downwards, and enters the centre of a vortex ring; seen in a vertical section, this ring appears as a pair of counter-rotating vortices, with a stopping vortex on the left and a starting vortex on the right. The axis of the vortex ring is nearly vertical, and the ring itself is horizontal: this is the 'horizontal vortex ring'. The air only rises slightly in front of the butterfly's body.

Pronation continues in the second frame (Fig. 9.32, 2), as the anterior edge of the wing moves backwards. The motion of the leading-edge vortex becomes more intense. A new vortex forms in the cavity of the outer edge of the forewings. It rotates anticlockwise towards the stopping vortex, and both vortices become part of a new vortex ring which is connected to the horizontal one by the old stopping vortex. The axis of this ring is horizontal and the ring is vertical: this is the 'vertical vortex ring'. The air which passes through the gap between the wings moves along the axis of the vertical vortex ring. As soon as it forms, the vertical vortex ring begins to move backwards.

In the third frame (Fig. 9.32, 3) the forewings are completely pronated. The outer edges of the hindwings come together to form a tunnel. Air pours through it, moving the vertical ring backwards, and the starting vortex begins to dissipate. In the next frame (Fig. 9.32, 4), the forewings have shifted backwards. The current through the tunnel intensifies, and the vertical ring continues to move backwards while at the same time gradually dissipating. The old starting vortex breaks down completely, and a new starting vortex is formed in the cavity of the outer edge of the forewing. It is important to note that the new starting vortex has no connection with the vertical vortex ring, which continues to move backwards.

The fifth frame (Fig. 9.32, 5) illustrates the beginning of rapid lowering of the wings. When the wings are completely pronated, they stop for a while, and only then does the downstroke begin. During this period, the air between the closed outer edges of the hindwings continues to move, and the new starting vortex increases its rate of rotation. The anterior edges of the forewings leave the sagittal plane, and consequently the leading edge vortex becomes invisible. The vertical vortex ring moves still further backwards and continues to dissipate.

In the next two frames (Fig. 9.32, 6 and 7) the wings are moving down, and the airflow velocity above the insect body increases. The starting vortex becomes wider, and shifts downwards and backwards. The vertical vortex ring dissipates completely. By the end of the downstroke (Fig. 9.32, 8), the starting vortex has shifted downwards and slightly backwards. The airflow velocity above the insect's body continues to increase. At this phase of the stroke, one can see the lines of airflow over the body and wings, as in the classical case of flow around an aerofoil.

When the wings reach the bottom of the stroke (Fig. 9.32, 9), a stopping vortex begins to form near the edges of the hindwings. The turning of the wings in preparation for the upstroke is accompanied by strong wing supination (Fig. 9.32, 10). The stopping vortex continues to increase and, together with the starting vortex, it forms a horizontal vortex ring. At this stage, the flow velocity above the body is still high (Fig. 9.32, 10 and 11). When the strongly supinated wings move upwards (Fig. 9.32, 11), the body becomes more vertical and the airflow points almost downwards. This completes the formation of the horizontal vortex ring. The last frame (Fig. 9.32, 12) shows the final phase of the upstroke; the leading edges of the right and left wings are about to meet. Passing through the narrowing gap between the wings, the air flow turns abruptly down through the hole in the vortex ring.

Each wingstroke therefore produces two vortex rings: one during the downstroke, and the other during the upstroke. They are coupled by the stopping vortex and form an L-shaped vortex structure that is thrown backwards at the top of the stroke.

9.4.2.2 A migratory flight.

One complete stroke cycle lasts about 0.1 s, and includes the same phases as the feeding flight. The main difference is that the wings produce a strong clap both at the top and at the bottom of the stroke. The stroke angle amplitude reaches 180°, and the wingbeat frequency is 10–13 Hz. Unlike in feeding flight, the body angle relative to the horizontal reaches its maximum value by the end of downstroke. It is interesting to note that in both feeding and migration flights the mean speed of the forewing tip in the stroke cycle is the same, approximately 2 m s^{-1}. However, the vortex formation pattern is different. At the top of the stroke, a butterfly in migration flight claps its wings, their dorsal surfaces being pressed firmly against each other. At the beginning of pronation, the vertical vortex ring moves quickly backwards. At the bottom of the stroke the wings shed the horizontal vortex ring downwards and backwards. Unlike in feeding flight, in which the horizontal ring is coupled to the vertical one and is therefore retained near the edges of the hind-wings (Fig. 9.32, 9–12), in migration flight the clap at the bottom of the stroke is accompanied by shedding of the vortex ring. Having left the wings, the horizontal vortex ring moves down-wards and backwards—the ring would move in the same direction during feeding flight if it were released at the bottom of the stroke (Fig. 9.32, 10). On cinefilm one can clearly see how the stopping vortex is formed. At the beginning of the upstroke, after the vortex ring has left the wings, a new starting vortex is created. A similar vortex ring was recorded by Ellington (1980) during take-off by the cabbage white. At the end of the first downstroke, the wings clap together, throwing a large-cored vortex ring downwards. Thus during migration flight the butterfly throws off vortex rings at both upper and lower points of the stroke.

9.4.2.3 Force production during flapping flight.

The data obtained by the observations described above enable us to estimate the dynamics of the forces acting on the butterfly during the stroke cycle. During both the upstroke and the downstroke the air is accelerated and is thrown downwards and backwards. The impulse received by the insect is therefore directed for-wards and upwards. The rate of change of the trailing and/or shed vortices is related to force generation, and this can be deduced from the cinefilm analysis. From the classical mechanics of incompressible fluids, conservation of flow demands that:

$$V_1 S_1 = V_2 S_2,$$

$$V_2 = V_1 \frac{S_1}{S_2},$$

where V_1 is the speed, normal to the cross section, of a tubular flow just in front of the butterfly's wing, and S_1 is the cross-sectional area of this tubular flow; V_2 and S_2 are the equivalent variables for the same tubular flow immediately behind the butterfly's wing. The phasing of force generation can thus be esti-mated with some degree of confidence from the relative velocities of the airflows just in front of and just behind the butterfly's wings and from their directions (Fig. 9.32). Further information can be obtained by identifying small changes in the pitch of the insect during the stroke cycle.

The analysis above shows that the velocity of air accelerated by the butterfly's wings reaches its maximum by the middle of the downstroke (Fig. 9.32, 7). At that point, the velocity of the airflow immediately behind the wings is three times that of the undisturbed flow. During the downstroke (Fig. 9.32, 5–8), a vortex ring clos-ing on the wings is formed. From the smoke streams, it can be seen that its hind part is a starting vortex. Its front part forms a circulatory flow around the wings which is responsible for the difference in flow velocities above and below the wings. The force impulse received by the butterfly is directed forwards and upwards. Hence the nature of force generation during the downstroke may be explained by quasi-steady aerofoil action.

At the bottom of the stroke, the wings shed the horizontal vortex ring by means of the stopping vortex (Fig. 9.32, 9). Even after it has been shed, this ring continues to play an important role in the accelerating flow above the butterfly's body (Fig. 9.32, 10–12), which is why the butterfly creates useful forces even when the wings stop momentarily at the bottom of the stroke (Fig. 9.32, 9).

During the upstroke negative circulation is created around the wings; this is shown by the formation of a stopping vortex behind the elevating wings, and it also constitutes a starting vortex for the upstroke (Fig. 9.32, 10–12). However, the negative circulation makes a negligible contribution to force production compared with the powerful current above the body which is created by the downstroke ring. Owing to the influence of this ring on the airflow, the force acting on the insect during the upstroke is directed forwards and upwards (Fig. 9.32, 10–12). Hence the nature of force generation during the upstroke (and also during supination) should be explained not by quasi-steady aerofoil action but by unsteady aerofoil action.

At the beginning of the clap (the near-clap for a feeding flight) the vertical vortex ring formed during the upstroke is shed (Fig. 9.32, 1). As it moves away from the wings (Fig. 9.32, 2–5), the horizontal vortex ring no longer accelerates the airflow above the insect. The velocity of the vertical vortex ring moving backwards is 1.5 times greater than that of the undisturbed air. As the vortex ring moves away from the butterfly, the hindwings form a tunnel through which a small amount of air is pushed backwards (Fig. 9.32, 2–4). A similar tunnel has been observed in the flight of butterflies of another family (Bocharova-Messner and Aksyuk 1981), and it was postulated that this tunnel played an important role in jet propulsion of the butterfly. It is not clear, however, if the backwards movement of the vertical vortex ring is caused by the air being pushed out of the tunnel or by some other force. The vertical vortex ring begins moving backwards (Fig. 9.32, 1) before the tunnel is formed (Fig. 9.32, 2), so it seems likely that during wing pronation the butterfly merely throws the vertical vortex ring backwards. If that is the case, the near-clap must create the jet motion, as Ellington (1984) predicted on theoretical grounds: 'If we neglect any circulation

remaining around the wings from the upstroke for a moment, then the clap would create the jet motion . . .' The flow visualization experiments show that the near-clap does produce the jet motion, although its precise mechanism is different from that predicted by Ellington. The air circulation which remains around the wings after the upstroke is important for creating the jet motion. Indeed, the vorticity shed from the leading edges of the wings as they clap rolls up onto the inner wall of the vertical vortex ring, directing the airflow backwards (Fig. 9.32, 2–4). Hence it is the near-clap which sheds the vertical vortex ring from the wings, and the subsequent pronation which throws it backwards (Fig. 9.33). This motion corresponds to a brief jet of air backwards which does not seem to play any significant role in butterfly movement. On the contrary, when butterfly wings push the vortex ring backwards, the butterfly must experience a forward reaction to this motion (Fig. 9.32, 2–4). The 'reactive force' acts on the butterfly's body until the starting vortex of the downstroke is formed (Fig. 9.32, 4), and the cycle is repeated.

The time sequence of force production during migration flight is slightly different from that described above. At the end of the downstroke the butterfly generates lift by casting the vortex ring downwards, while at the upper point of the stroke the butterfly casts the vortex ring backwards and hence generates thrust. Clearly, the butterfly may increase its flying speed by

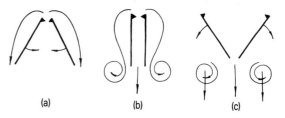

(a)　　　　(b)　　　　(c)

Fig. 9.33 Flow patterns during the near-clap in the feeding flight of *Inachis io*, view from above: (a) clap of the leading edges of the wings, (b) vortex shedding, and (c) pronation of the wings followed by throwing the vortex ring off backwards. Arrows indicate the direction of movement of the wings and the airflow.

casting the downstroke vortex ring backwards and downwards.

9.4.3 Flight peculiarities of insects with low-frequency-type kinematics

The wing kinematics of broad-winged butterflies are of a low-frequency type: the broad fore- and hindwings produce synchronous low-frequency and large-amplitude flaps along a trajectory which is vertical relative to the longitudinal axis of the body. The insect covers a relatively long distance with each wing flap, and periodical changes in altitude and direction make the flight fluttery. The flight speed is high. Jerks carried out with the wings lifted above the body are very characteristic. As useful force is being continuously generated the butterfly can make fast forward jerks without losing altitude. When the butterfly continues to move upwards by inertia (Fig. 9.32, 1), the ring is thrown backwards, providing additional forward thrust (Fig. 9.32, 2–5).

The inclusion of gliding into the general flight profile makes the flight trajectory unpredictably complicated. Their ability to glide enables broad-winged butterflies to be magnificent migrants capable of covering considerable distances.

The loss of altitude which can be observed periodically in free feeding flight is accomplished by a temporary halt in the stroke cycle. The adaptive importance of such behaviour is apparently related to avoiding potential enemies (such as insectivorous birds). However, this behaviour is absent from fast, steady migratory and searching flights.

9.5 Conclusion

The appearance of a coupling mechanism between the fore- and hindwings was an important event in the evolution of the winged insects, allowing them to increase their wingbeat frequency. The principal consequence of this increase in wingbeat frequency was the development of a new mechanism of closing the vortex rings on the wings, as a result of which manoeuvrability increased sharply, as the wings on opposite sides were transformed into independent vortex ring generators. The main load in wing operation passed to the median notal wing process, which became differentiated into a median and a posteromedian processes. Scutellar arms were formed, and finer control of wing movement became possible by use of the lateropostnota. Wing deformation during strokes changed from supinational bending to the rapid passage of a deformational wave over the wing. For the first time in the evolution of insects the wing plane began to be strongly deformed during pronation.

Wing folding improved during the same period. A complex system of levers, which developed at the wing base, ensured semiautomatic roof-like folding, this system operating differently in the fore- and hindwings.

The best butterfly wing shapes for gliding are those of the *Heliconius*-type. *Papilio*-type butterflies, which are just as good at gliding as the *Heliconius*-type, are distinguished by a significantly lower wing aspect ratio and a complex outline of the terminal margin of the hindwings. It is suggested that in all *Papilio*-type butterflies the flow becomes divided into tip and root parts, and the wake consists of the two pairs of vortex braids. Significant responsibility for the formation of the inner pair of vortex braids and, consequently, for the decrease in drag falls on the tails of the hindwings. Butterflies with a low aspect ratio but which lack the tails achieve the improvement of aerodynamic characteristics of the wings by increased roughness of the lower wing surface, by means of scales and hairs. During the rare flaps of the broad wings the vortex wake, which consists of discrete large-diameter vortex rings, is left behind.

10

Looking into the past:
the process of evolution, and insect wing apparatus

Given the basic animal postulate—an economy predatory upon plants and other animals—the first necessary corollary of this is LOCOMOTION.

Weston LaBarre

The appearance of the wing apparatus was a key point in the evolution of insects—an animal class rich in species and of great theoretical and practical importance. In the course of over 300 million years of development the wing apparatus underwent significant changes that affected the structure of the wings, axillary apparatus, exoskeleton, and musculature as well as the mechanism of wing kinematics and deformation, and also influenced flight aerodynamics—the principles underlying the creation of the forces necessary for locomotion in the air. At a certain stage of evolution, it became possible to start storing the energy derived from oscillation by means of resilient deformation of the thorax.

The historical development of flight-enabling systems proceeded as a chain of interconnected and interrelated events that directed selection towards increases in flight efficiency. The principal selective advantages were increases in the economy, manoeuvrability, and velocity of flight. The first of these led to the convergence in the development of gliding flight in the anisopterous Odonata and Rhopalocera, and the second resulted in independent movement of the wings of the right and left sides of the body in Odonata and Diptera, a capacity that arose independently in these two groups and was based on different mechanisms. The increase in flight velocity, which, given constant body size, could be achieved by increasing the wingbeat frequency, occurred in parallel during the evolution of several different lines of winged insects.

10.1 Types of wing kinematics

Growth in wing pair synchronicity and increased wingbeat frequency played a central role in the evolution of insect wing kinematics. Wingbeat amplitude was also reduced and the trajectory of wing movement became straightened, the angle of wing rotation around its longitudinal axis being increased at the extremes of the trajectory.

The various phase relationships between the movements of the wing pairs and stroke parameter combinations define a range of different modes of action of the wing apparatus. For his classification of insect wing apparatus, Rohdendorf (1949) primarily made use of wing structure and the extent of development of the pterothorax musculature, regretting that 'the main problem to be solved—the analysis of the insect wing stroke—is still far from its final solution'. Yet the most profound differences between in insect wing apparatus occur in wing kinematics. The evolution of insect flight consists mainly of the evolution of insect wing kinematics (Fig. 10.1). The nine principal types of wing kinematics of extinct and recent insects are described below: one further type was that of the Palaeodictyoptera, in which both wing pairs worked synchronously at low stroke angles, alternating with gliding.

1. *Eosynchronous kinematics.* The wings move along a steep trajectory at a high stroke angle and low frequency. Among recent insects this type is characteristic of mayflies which seem to have preserved the original wing pairs movement synchrony. This mode of flight eventually resulted in complete or partial reduction of the hindwings in several mayfly groups. The hindwings have lost their independence and

are pulled along by the forewings during flight. However, in spite of a superficial similarity with functional two-wingedness, wingbeat frequency has not increased significantly, although as usual it depends on body size.

2. *Bi-motor kinematics.* Each of the four wings moves independently, and the wing pairs work at any phase relationship, the most frequent phase shifts, however, being a quarter or a half the cycle. This is typical of modern dragonflies, and may also have occurred in some Megasecoptera. Anisopterous Odonata have extremely small stroke angles, whereas zygopterous Odonata have slightly larger stroke angles and a significant freedom of stroke plane orientation. The wingbeat frequency is rather low.

3. *Quasi-synchronous kinematics.* The wings move slightly out of synchrony, at equal and

Fig. 10.1 Basic phylogenetic tree of winged insects as represented by the historical development of their wing kinematics. In the process of evolution, along with an occasional use of gliding (b), there was a sequence of the following types of wing kinematics: (a) eosynchronous, (c) bimotor, (d) quasi-synchronous, (e) hind-motor and true hind-motor, (f) functionally four-winged, and (g) functionally two-winged and (h) the eventual appearance of asynchronous flight muscles. The numbers 1–4 correspond to four main branching points of the phylogenetic tree: 1, occupation of new adaptive zones (arboreal, exposed, and ground-level) by representatives of three main evolutionary lines; 2, divergence of the Polyneoptera from the Paraneoptera and the Oligoneoptera at the time when functionally four-winged flight appeared and the Polyneoptera began specializing for a concealed lifestyle; 3, divergence of the Paraneoptera from the Oligoneoptera, at the time when feeding on liquid by means of sucking mouthparts made insects spend a long time in one spot and hence improve their wing folding. Unlike the Paraneoptera, flight technique was still improving in the Oligoneoptera; 4, divergence within the holometabolans which led, in some lineages, to the development of extremely powerful, perfect flight.

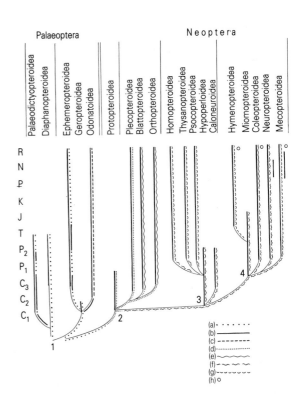

large stroke angles. The trajectories of the fore- and hindwing have a complex shape and lie at different angles to the horizontal. The wingbeat frequency is rather low. This form originated from eosynchronous kinematics, when the forewings began to function partially as a cover; it still occurs in stoneflies.

4. *Hind-motor kinematics.* The stroke angle of the hindwings is significantly greater than that of the forewings which function as a cover or serve other functions (crickets), and are also employed in flight control. The wingbeat frequency is low. The dorsal surface hindwing clap is usually present. This form is characteristic of all Polyneoptera, except stoneflies.

5. *Functionally four-winged kinematics.* The wing pairs move downwards synchronously and upwards asynchronously. The hindwings beat at a greater stroke angle than the forewings. The wingbeat frequency is low. The dorsal surface forewing clap is usually present. This form probably originated from quasi-synchronous kinematics as a result of a slight shortening of the lower part of the forewing trajectories, and occurred in the ancestors of the Para- and Oligoneoptera. It is retained in a number of modern insect orders (Fig. 10.1).

6. *Functionally two-winged kinematics.* The coupled fore- and hindwing pairs move strictly in synchrony, at a moderately high wingbeat frequency. The stroke angle and the stroke plane angle relative to the body axis are lower than those of functionally four-winged insects. The stroke plane angle may change significantly. This form is typical of the Homoptera plus Heteroptera and of some Lepidoptera and Trichoptera. Depending on the role played by either the fore- or the hindwing pair in the generation of aerodynamic forces, this form may be divided into two subtypes, forewing kinematics (some Lepidoptera, Homoptera, and a few Heteroptera), and hindwing kinematics, in which the forewing pair serves partly as a cover but nevertheless continues to lead the hindwing pair during flight, while the hindwing pair contributes to the formation of aerodynamic forces (Trichoptera, some Lepidoptera, Homoptera, and most Heteroptera).

7. *Low-frequency kinematics.* The broad fore- and hindwings move in synchrony with an extremely low wingbeat frequency and at a high stroke angle along a vertical trajectory relative to the longitudinal axis of the body. This form is found in most Rhopalocera and in several large, broad-winged moths, mostly Saturniidae.

8. *True hind-motor kinematics.* Aerodynamic forces are generated only by the hindwings which flap with a large stroke angle, and with claps at the highest and lowest points of the stroke. This form is characteristic of the Coleoptera. The wingbeat frequency is strongly dependent on body size.

9. *High-frequency kinematics.* The wings move along a sloping trajectory at a relatively low stroke angle and a high wingbeat frequency. The stroke plane position can vary significantly relative to the longitudinal body axis, depending on the mode of flight. This form is seen in the Hymenoptera and Diptera. Features like the presence of only one wing pair and a greater ability of the wings to rotate around their longitudinal axes separate a special subtype from the Hymenoptera—the anatomically two-winged form. This subtype is unique to the Diptera.

10.2 The correlation between wing supination and wing pronation

The first winged insects apparently inherited from their ancestors some basic properties of the wing plate. First, the wing plate bends longitudinally more easily upwards than downwards; second, transverse bending occurs downwards only. These basic properties clearly determined both the most primitive pattern of wing deformation and the whole sequence of their evolutionary stages.

Insect wing deformation may be caused by

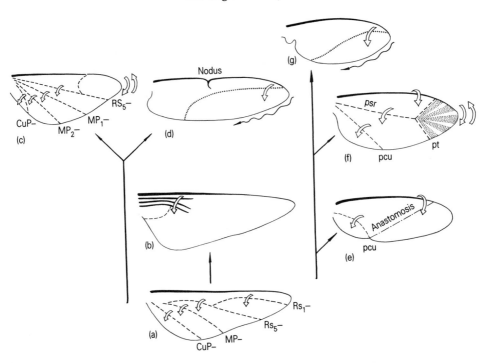

Fig. 10.2 Diagrammatic representation of the evolution of supinational deformation. The most primitive type of supinational deformation occurred typically in (a) *Eugeropteron*, which gave rise to that of (b) the Palaeodictyoptera, and also to that of (c) the Ephemeroptera, and (d) the anisopterous Odonata. The other branch follows the sequence from (e) supinational twisting via (f) supinational bending to (g) the tip-to-base deformational wave. Open arrows indicate the direction of deflection of certain parts of the wing during supination.

aerodynamic, inertial, elastic, and muscular forces. The strongest deformation of insect wings is induced by the forces of inertia at the lowest point of the downstroke, when the wings begin to rise and supinate. The types of wing deformation, which occur during supination, which are called supinational deformation, were considered in the chapters dealing with the flight peculiarities of the different groups of insects. The stages in, and sequence of, the evolution of the various forms of supinational deformation are shown in Fig. 10.2. The most primitive form, which was inherited from *Eugeropteron*, is represented by mayfly wings. The supinational deformation of palaeodictyopteran wings was of a distinct, rather peculiar type. The acme of the development of supinational deformation is a wave passing down wings whose veins are provided

with various hinges and whose membranes are soft and flexible. This type of supinational deformation arose independently in several different phylogenetic branches (Fig. 10.2). The membrane softness and flexibility of dipterous wings are achieved by liberation of the membrane from veins, while in dragonflies these were achieved by the attachment of the numerous transverse veins to the longitudinal ones by hinges.

The most widespread type of supinational deformation is supinational bending (Fig. 10.2f). This required the development of a semi-circular or pointed tip deflection line on the wing, as well as a radial sector fold. In this case, inertia from the downward wing movement causes elastic torsion in the vertical hinge region as well as a sharp wing tip deflection downwards. The backwards movement of the wing which is induced

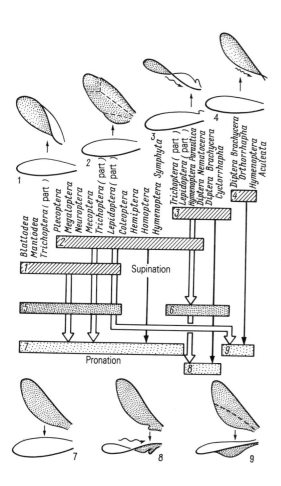

by elastic forces is accompanied (apart from the inevitable supinational twisting as the leading edge moves more quickly than the trailing edge) by straightening of the wing tip. The return of the tip to its original state allows the wing blade to bend longitudinally along the radial sector fold, ensuring additional supination of the wing. The wing then moves upwards with the tip region bent longitudinally, and sometimes with the greater part of the remigium supinated along *psr*.

Owing to the elasticity of wing veins each supinational deformation initiates a corresponding pronational deformation (Fig. 10.3). Thus supinational twisting causes an opposite phenomenon—pronational untwisting, during which convex bending of the rear edge of the wing shifts apically and the wing plane straightens up. Further development of this deformation leads, in turn, to pronational twisting, in which the convex bending of the leading edge shifts towards the base. The screw-like torsion which reduces supinational twisting often follows immediately after the supinational twisting itself, a phenomenon which occurs, for instance, in stoneflies (see Fig. 7.5, 2 and 3, p.120). In other cases pronational untwisting of the wing blade may occur in the last third of the ascending branch of the wing tip trajectory (Blattodea, Megaloptera, Neuroptera, etc.). In the skipper *Thymelicus lineola*, pronational untwisting of the wings begins as the wings approach the highest point of the stroke, so that when the wings begin going downwards, their tips are already pronated; in this case muscular effort at the wing bases is enough to achieve complete pronational turning of the wings. In broad-winged butterflies, supinational bending of the wings also causes pronational bending of the wings at the top of the stroke by means of elastic deformational forces. In the Diptera and Hymenoptera the correlation between supinational and pronational deformation is even more pronounced.

Wing pronation at the beginning of the downstroke occurs as a result of basalar muscle contraction. In the most simple case, the costal margin of the wing is simply turned down like a rigid blade which does not deform (Fig. 10.3, 7). This phenomenon is seen in relatively primitive

Fig. 10.3 The relationship between various types of wing deformation at different sections of the wing trajectory, by insect orders. Supinational deformation: 1, supinational twisting; 2, supinational bending; 3, tip-to-base rear margin wave; 4, supinational rotation. Pronational deformation: 5, pronational untwisting; 6, base-to-tip rear margin wave; 7, pronational turning; 8, tip-to-base rear margin wave; 9, pronational bending. Sections of the wing trajectory: 1–4, bottom; 5 and 6, ascending branch; 7–9, top. Thin arrows show the relationship between the stages of pronational deformation and the corresponding supinational deformation; open arrows show how elastic forces are used on the ascending branch of the trajectory. The lower surface of the wing is shaded.

insects: cockroaches, mantids, stoneflies, etc. A more complex type of wing pronation involves ventral longitudinal bending of the wing (Fig. 10.3, 9). This appears only in some Lepidoptera, the Diptera Orthorrhapha, and Syrphidae. In some insects, such as the Hymenoptera Parasitica and the Diptera Nematocera, acute pronational bending only involves a narrow wing tip region, starting from which a wave-type deformation passes towards the body, resulting in pronation of the entire wing blade (Fig. 10.3, 8). Pronational waves always spread faster than supinational ones.

The most severe wing deformation takes place when the wing is supinated by inertial forces. In the course of evolution gradual supination appeared first, and was later replaced by torsion, then followed by tip deflection, and finally by the wave-type deformation running from the tip to the base of the wing. The peak of development of supinational deformation was reached by the Diptera Orthorrhapha, Syrphidae, and Hymenoptera Aculeata. No strong bending of the wing blade occurs in these groups.

Pronational deformation, by contrast, is characterized by only slight bending of the wing blade in its most primitive state. Wing blade bending becomes intensified with the appearance of additional longitudinal folds and the re-distribution of rigid fields, as can be observed in the broad-winged Lepidoptera and in some Diptera. Thus, taking into account the need for supinational deformation, a wing structure was formed in the course of evolution which was capable of ensuring, in its most advanced state, strong wing deformation during pronation.

10.3 The role of deformation in the evolution of wing structure

The wing has to meet the following requirements as a locomotive structure:

(1) to sustain aerodynamic pressure, i.e. to be strong and flexible at the same time;
(2) to resist the forces of inertia, i.e. to be regularly deformed during the stroke cycle;
(3) to interact with the surrounding air currents.

Mechanical strength is provided by the system of veins, deformability is caused by the system of folds, and wing texturing provides the interaction with the airflow.

The general venation pattern changed in the course of evolution towards more economical means ensuring of mechanical strength. The most primitive type of venation, the archedictyon, is characterized by parallel longitudinal veins joined by a dense network of transverse veins which form numerous cells of various shapes. The regular arrangement of transverse veins between the parallel longitudinal ones results in 'polyneural venation'. In more advanced forms of polyneural venation the longitudinal veins are thicker than the transverse ones. Reduction of the transverse veins leads to the formation of 'oligoneural venation'.

A higher level of wing framework modification is marked by cell venation in which the cells are formed not only by the transverse veins, but also by curved longitudinal veins. Folds and cells may be regarded as antagonistic: the folds increase the wing's flexibility perpendicular to the veins, whereas cells reduce it. A wing made up of small cells is unable to bend sharply but, on the other hand, its general strength, combining rigidity with flexibility, is increased. A thick network of transverse veins is not always unadaptive. In some cases, the requirements of the mechanical properties of the wing can be better met with a cell structure close to the archedictyon. For instance, the wings of a green lacewing bend strongly but relatively slowly when manoeuvring. These conditions are best satisfied by wings with a dense cell network and poorly pronounced folds.

The main thrust in the evolution of wing venation is associated with costalization of the wing—with an increase in the rigidity of the leading edge. This has been achieved in a variety of different ways. Most frequently the anterior longitudinal veins have shifted to a submarginal position, accompanied by thickening of the veins. In addition, the posterior veins bend and, before reaching the rear edge, they either turn towards the leading edge or else become gradually thinner and thinner until they disappear.

There has also been a trend towards levelling of the macrorelief. The pronounced corrugation typical of relatively primitive insects was replaced by a comparatively even surface above which the veins project slightly. In general, the wing surface texture is produced by a combination of the folds and waves of the membrane, emphasized by the microrelief. The wing membrane in the Ephemeroptera, Odonata, and Neuroptera is bald, whereas in representatives of higher insect orders the wings are covered with an uninterrupted layer of structures which are often even arranged in two tiers (Bocharova-Messner 1979). Since the microstructures are present on both the dorsal and the ventral sides of the wing, the overall thickness of the wing increases fourfold. Hence the inertial forces which affect the wing during a stroke are also four times stronger. As a result, in the first place, the longitudinal veins grow thicker because of the increase in load and, in the second place, the various folds along which the wing is bent begin to play a more important role. These events lead to increased complexity of the general wing structure, the appearance and improvement of hinge-like units, and differentiation of functional regions.

The arrangement of functional elements on the insect wings is governed by certain strict rules. The main mechanical axis of the wing always runs close to the leading edge of the wing blade, the lines of pronation and supination are shifted backwards, and downward tip deflection occurs along the angular line associated with the radial sector fold. The vein tips approach the wing edge in parallel rows, thus providing the prerequisite for the formation of grooves. The radial sector fold and the cubital fold exert especially powerful influences on the wing structure. These folds divide the wing into three relatively independent fields, each having its own system of supporting elements. The anterior field is strengthened by the main mechanical axis, the posterior one by the postcubitus, and the median one, which contains the medial and cubital

stems, can move freely relative to the anterior functional field.

Hinges form at the points where the veins and the folds cross. The hinges function as resilient copulae, ensuring strength and flexibility of the wing; they also govern the direction of deformation. In some cases, as veins cross a fold, their changes in morphology are undetectable; this is especially true of the main veins. This situation often occurs when a fold reaches a vein at a very sharp angle, almost tangentially. In this case the deformation along the fold leads to twisting rather than to bending of the vein; the vein is not constricted, so that a hinge is unnecessary. The most important hinges are arranged along the folds *psr* and *pcu*. Also noteworthy is the hinge situated on the radius in the subcostal vein tip region which is associated with various types of transverse deformation of the wing.

The position of deformation lines often governs the pattern of venation. Both folds and other deformation lines determine the 'x' shaped arrangement of stonefly veins and the arrangement of the longitudinal veins in the cubital field in mayflies (Fig. 10.4a and b). In alderflies and caddisflies, at the end of the upstroke, convex ridges are formed; in *Sialis* and *Phryganea* these ridges still conform to the pattern of venation and are irregularly arranged (Fig. 10.4c and d), but in the more advanced *Neuronia* the tip of the radial vein is curved and the discoidal cell shortened and rearranged to produce the prerequisite for the formation of a rounded ridge (Fig. 10.4e). Rearrangement of the cubital vein system in caddisflies, when compared with that of alderflies (Fig. 10.4f), is associated with the curious manner in which the cubital fold passes through the corresponding region (Figs. 10.4g and h). Transverse grooves which appear on the clavus of stoneflies during strokes lead to the veins bending and becoming more slender (Brodsky 1982). In hymenopterans, longitudinal grooves are formed on the clavus of forewings when they are coupled with the hindwings, resulting in the formation of the anal loop (Fig. 10.4i). Similar grooves must have appeared in the Homoptera, Trichoptera, and some Lepidoptera whose wings are coupled and whose forewings have anal loops (Fig. 10.4j and k). The same phenomena presumably caused the formation of anal loops in extinct insects (Fig. 10.4l and m).

In some cases, the wing deformation has strongly affected wing morphology. Thus, in the extinct Glosselytrodea the radial sector fold played an important role during flight. This led to radical reconstruction of the entire wing. As a result, a long membranous field without any transverse vein formed along the whole length of *psr*, and the wing became approximately symmetrical (Fig. 10.4n). The radial sector fold may also have given rise to the unusual arrangement of veins on the wing base of some extinct Grylloblattida (Fig. 10.4o).

The origin of transverse wing deformation was associated with the increase in inertial forces as the wing tip began to move faster owing to the elongation of the wing, or an increase in wingbeat frequency. Palaeozoic insects already had structures which facilitated and localized transverse deformation—hinges on the main mechanical axis of the wing. These hinges arose independently in different groups, which

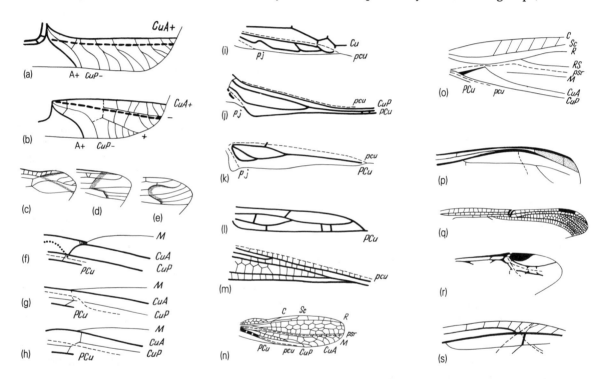

Fig. 10.4 Structural details of the wings of extinct and extant insects. (a) and (b) the groove (bold broken line) that appears in the cubital field of the forewing in mayflies during wingstrokes: (a) *Ephemera vulgata*, (b) *Ephoron virgo*; (c)–(e) ridges at the wing tip (shaded): (c) the alderfly *Sialis morio*, (d) and (e) the caddisflies *Phryganea grandis* and *Neuronia ruficrus*; (f)–(h) basal region of the mediocubital field of the forewing (dotted line shows the site of wing deformation): (f) the alderfly *Sialis morio*, (g) and (h) the caddisflies *Phryganea grandis* and *Neuronia ruficrus* respectively; (i–m) the anal loop: (i) the hymenopteran *Macroxyela* sp., (j) the fulgorid planthopper *Fulgora maculata*, (k) the caddisfly *Rhyacophila* sp., (l) the palaeozoic 'caddisfly' *Permomerope karaungirica*, (m) the glosselytrodean *Eoglosselytrum kaltanicum*; (n) and (o) system of folds: (n) the glosselytrodean *Archoglossopterum shoricum*, (o) the grylloblattid *Stegopterum hirtum*; (p–s) tip hinges: (p) the megasecopterid *Siberohymen asiaticus*, (q) the dragonfly *Gomphus descriptus*, (r) the hymenopteran *Pteronidea ribesii*, (s) the stonefly *Isogenus nubecula*.

is clearly visible by comparing these structures in the Odonata, Plecoptera, Hymenoptera, and Megasecoptera (Fig. 10.4p–s). An interesting case occurs in the Palaeozoic Blattinopseida: the wings of these insects had an arciform deflection line which crossed the longitudinal veins, and there were longitudinal grooves between neighbouring veins in the deflecting part of the wing. These grooves were permanent structures which originated from temporary furrows, normally formed during straightening of the wing tip.

The examples discussed above make it obvious that deformations exerted a powerful influence on the formation of the insect wing structure.

These deformations act as follows: first a change in the function of a certain wing part occurs and this is accompanied by the development of folds and other deformational elements, which in turn lead to changes in the venation pattern (Fig. 10.4a–h)

The three main overall trends in the evolution of wing structure are as follows:

(1) ensuring the necessary mechanical strength of the wing by means of the most energy-sparing method;
(2) the development of a system of folds;
(3) the development of microrelief.

10.4 Increase in efficiency of effort transmission from muscles to wings

Although the wing apparatus functions as a whole, each of its components is principally influenced by local factors. Thus, the structure of the wing is determined by its mode of supinational deformation in lower insect orders, and the mode of pronational deformation in higher orders. The structure of the axillary apparatus depends, first of all, on specific kinematic features of the wings and on specific wing-folding habits. The greatest stability is found in the wing muscles, whose make-up is determined mainly by phylogeny. Flight aerodynamics is the strongest influence on wing shape which in turn controls the nature of supinational deformation and therefore the wing structure. The combination of all these factors, namely aerodynamics, kinematics, deformation, and phylogeny, defines which type of wing apparatus organization occurs in a given insect group. Wing structure appears to depend on the largest number of factors, and hence there is a very broad variety of wing designs. Study of the pterothorax structure is therefore of the greatest importance for the elucidation of phylogenetic relations among large insect taxa.

10.4.1 The skeleton and musculature

In the first winged insects, the tergum of the wing-bearing segment seems to have been equipped with all the structures necessary for flapping; that is, a complete set of sutures, sulci,

and ridges. The concentration of the superior ends of the dorsoventral muscles in the most crucial points of the notum became a common trend in the evolution of the pterothorax. Furthermore, in connection with enhancement of the flight function, some muscles split into bundles (for example, *t12* in Homoptera Auchenorrhyncha), which lead to changes in the locations and areas of muscle attachment on the inner surface of the notum. This led to the isolation of notal regions that bear particular loads during flight. The diffuse distribution of superior muscle ends around the notum is a primitive feature. Another common trend in the evolution of the pterothorax was the reduction of synergistic muscles. This oligomerization process ran in parallel to the growth in volume of the remaining muscles. An extreme manifestation of this trend can be seen in the Hymenoptera Aculeata.

The increase in the role of the dorsal longitudinal muscles in wing movement was accompanied by elongation of the scutum and changes in the shape of the parapsidal sutures. The dorsal longitudinal muscles play the most important role in wing movement in the Ephemeroptera, Diptera Brachycera, and Hymenoptera Aculeata. In these insects the parapsidal sutures run in parallel.

The appearance of new sutures and other tergal structures is a rare event in the evolution of winged insects. It is evident that only a very significant change in the nature of wing movement, or, to be more precise, in the way in which the effort is transmitted from the contracting muscles to the wings can lead to the formation of new sutures and sulci. The transscutal fissure of the coleopteran metatergum was one of these. The development of this fissure made it possible for the nature of scutum distortion to be transformed so that the main load in wing operation falls not on the median notal wing process, as is the case in other Oligoneoptera, but on the anterior notal wing process. Similarly, the development of the transnotal sutures of the tergum in Diptera results from the increase in volume of the tergotrochanteral muscle, whereas in other insects this muscle does not play such an important role in the operation of the wing apparatus. The supplementary parapsidal sutures in higher Hymenoptera are absolutely unique; these were formed quite independently from a process of isolation of the regions where the dorsoventral muscles were attached to the notum.

The scutoscutellar suture should also be classified as a newly formed tergal suture, although it evidently appeared very early and its transformation process was closely associated with the fate of another—the oblique—suture. The two oblique sutures probably appeared at the same time as the recurrent scutoscutellar one, as the longitudinal dorsal musculature became involved in wing operation. These three sutures together form a triangular structure which focuses effort from contraction of muscle $t14$ on regions of the notal margin at the level of the anterior and median notal wing processes. With the formation of the scutoscutellar suture and consequently of hinges where it crosses the recurrent scutoscutellar suture, the mobility of the oblique sutures increased. The typical ephemeropteran notum structure was formed at this stage of suture development. In these insects, the oblique suture runs into the notal margin between the anterior and posterior notal wing processes. The median notal wing process is reduced. In the Polyneoptera, the anterior notal wing process is differentiated into anterior and anteromedian notal wing processes; the oblique suture runs into the notal margin

between the anteromedian notal wing process and the median one. Associated with this, in the Polyneoptera, the oblique suture is sometimes referred to as transscutal (Matsuda 1970). In the Paraneoptera, the oblique suture nears the notal margin at the level of the fused anterior and median notal wing processes. In the Oligoneoptera, whose median notal wing process is differentiated into the proper median and posteromedian, the oblique suture enters the notal margin between these processes and is not, therefore, homologous to the posterolateral scutal suture with which it is often confused (Matsuda 1970). We can therefore infer that the oblique sutures arose in the general tergum outline only once, by the involvement of the dorsal longitudinal musculature in wing operation. The tergal fissures developed primarily between the anterior and median notal wing processes in all insects. As can be seen, the oblique sutures play different roles in the mechanics of tergum movement in different groups of winged insects depending on the differentiation of the notal margin, whereas the role of the tergal fissures tends to decrease in the course of evolution. An exception to this rule are those groups in which transscutal sulci have developed based on the tergal fissures: the Orthoptera (in the metathorax), Odonata, and Hymenoptera.

Similar changes in function, occurring in parallel in unrelated groups, led to the formation of analagous structures. The development of the hind-motor state led to the formation of parallel structures such as the transscutal sulcus in the Orthoptera and the transscutal fissure in the Coleoptera. These structures have different origins but had a common outcome: focusing of the effort from contraction of the dorsal longitudinal muscles on the anterior notal wing processes. The formation of lateral parapsidal sutures in various groups can be regarded as a very simple example of parallelism. It is not fully clear why the Oligoneoptera lack lateral parapsidal sutures. One explanation may be that in this group the areas of attachment of the dorsoventral muscles closely adjoin the oblique sutures, leaving no space for the formation of supplementary sutures. In the Ephemeroptera the lateral parapsidal sutures externally delimit the areas of attachment of muscle *t–p5, 6* to the notum. In the Plecoptera these sutures separate the regions where the tergocoxal muscles and *t12* are attached. In the Homoptera Auchenorrhyncha the lateral parapsidal suture serves as a support for the powerful dorsal oblique muscle.

Striking evidence of parallelism in the evolution of the tergum is provided by the formation of scutellar arms in the Ephemeroptera, higher Hymenoptera, and Diptera. The scutellar arms of these insects have a similar structure and function but consist of incompletely homologous elements. In mayflies the scutellar arm is delimited by the oblique suture and the posterolateral scutal sulcus and carries the posterior notal wing process. In the lower Hymenoptera and Diptera the scutellar arm is delimited by the same suture and the same sulcus, but the median notal wing process is differentiated into median and posteromedian notal wing processes, the latter overlapping the top of the scutellar arm. In the higher Hymenoptera, the transscutal sulcus is formed in place of the tergal fissure, delimiting medially the scutellar arm which bears the median and posteromedian wing processes. In the higher Diptera, by contrast, the transscutal sulcus is not at all developed and the tergal fissure separates off the parascutum which bears the median notal wing process. Thus, the increase in scutellar arm mobility caused profound parallel changes in tergal structure in the Hymenoptera, Diptera, and to some extent in the Ephemeroptera. As a result of these changes, higher members of these orders sometimes show more similarity to each other than do the relatively primitive ones.

It is also necessary to consider the marked similarity in tergal structure between the Plecoptera and Megaloptera. The list of features shared by these insect groups (Adams 1958, Emelyanov 1977) can be added to by the virtually identical structure of the tergum of the mesothorax. The resemblance is so complete that it is sometimes hard to determine which of the two orders an insect belongs to using tergal structure. Were it not for the presence of the scutoscutellar suture in more advanced stoneflies (Fig. 10.5a), it might be assumed that muscle *t13* in Megaloptera is identical to muscle

t12 in Plecoptera. The marked similarity in structure of the pterothorax between the Plecoptera and Megaloptera, on the one hand emphasizes the parallelism in their development and, on the other hand, allows us to regard both groups as a kind of 'starting point' in the evolution of the Polyneoptera and Oligoneoptera respectively. Moreover, with regard to the organization of the wing apparatus, stoneflies may be regarded as the most generalized group among all present-day winged insects.

To be able to estimate the relationships between the main phylogenetic branches, it is necessary to summarize the apomorphic changes which occurred in the wing apparatus of the insects in these branches. Comparative analysis of the structure and function of the pterothorax of different orders of insects demonstrates the integrity of four phylogenetic lineages as follows: Ephemeropteroidea plus Odonatoidea, Polyneoptera, Paraneoptera, and Oligoneoptera. Each of these lineages has a common structure of which all extremes are variants. Equal development of the scutoscutellar and recurrent scutoscutellar sutures is typical in the Palaeoptera or at least in the Ephemeroptera (Fig. 10.5b), the scutoscutellar suture branches terminating at the recurrent scutoscutellar suture but not continuing further, as they do in Paraneoptera (Figs 10.5c and d). Divergence of the notal margins distally along the oblique suture (see Fig. 6.6, p.110) as it approaches the notal margin is also typical. The posterior notal wing process plays the main role in wing operation. This may be why the third axillary sclerites of mayflies and dragonflies are similarly shaped; the site of attachment of the muscle of third axillary sclerite to the tip of the pleural wing process is also identical (Fig. 10.6). In spite of all the differences in pterothorax organization between mayflies and dragonflies, it is evident that the further back in time we look, the closer together are the phylogenetic lines of these insects. In mayflies and dragonflies the pleural wing process is divided into two lobes, *PWP–1* and *PWP–2* (Fig. 10.6). *PWP–1* articulates with the basalar sclerite, and *PWP–2* serves as a support for the immobile end of muscle *t–p14*. In addition, the inferior layer of the basisubcostale articulates by means of a head-like hinge with

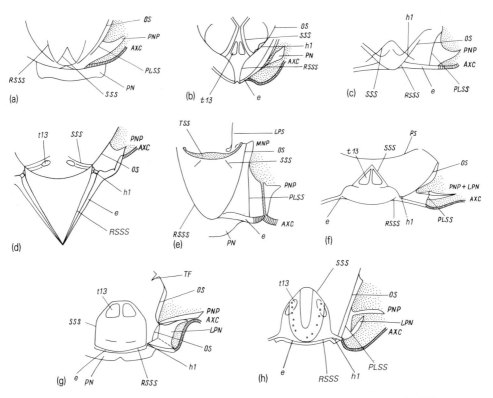

Fig. 10.5 Structure of the posterior part of the notum in representatives of different orders. (a) Mesonotum of the stonefly *Kamimuria luteicauda*, (b) mesonotum of the mayfly *Ephemera vulgata*, (c) mesonotum of the booklouse *Psococerastis gibbosus*, (d) mesonotum of the leafhopper *Aetalion reticulatum*, (e) metanotum of the bush cricket *Decticus albifrons*, (f) mesonotum of the alderfly *Sialis* sp., (g) mesonotum of the antlion *Acanthaclisis occitanica*, (h) mesonotum of the caddisfly *Phryganea bipunctata*.

the posterior lobe of the basalar sclerite; in dragonflies this head (Fig. 10.6a) is involved in the movement of the lower wall of the radial axillary plate, and in mayflies it forms a supplementary pivot (Fig. 10.6b) around which the wing twists during the downstroke. The tergum of the wing-bearing segment of the Meganeurida has a more archaic structure than that of the recent Odonata; the Meganeurida have many features in common with Ephemeroptera.

In the Polyneoptera, the recurrent scutoscutellar suture markedly predominates over the scutoscutellar suture (Fig. 10.5a and e); muscle *t20* is also present. The anterior notal wing

process has the principal role in wing operation. The oblique dorsal muscle is hypertrophied in stoneflies and cockroaches. As a result of the development of the hind-motor state in the Orthoptera, the transscutal sulcus has deepened, supplementary hinges have formed, and the scutoscutellar suture and the oblique dorsal muscle have become reduced (Fig. 10.5e).

The scutoscutellar and recurrent scutoscutellar sutures are equally typically developed in the Paraneoptera; the scutoscutellar suture branches run across the line joining the sutures (Fig. 10.5 c and d). Moreover, in these insects the oblique dorsal muscle is hypertrophied, the mechanism which locks the wings at rest is strengthened and complex, the anterior notal wing process is fused with the median one, and the posterior wing process is reduced.

The Oligoneoptera are characterized by shortening of the recurrent scutoscutellar suture (Fig. 10.5 f–h). Projection of the lateropostnotum on to the dorsal surface of the wing and the presence of a well-developed muscle *t13* are the more significant of a range of synapomorphies which are present. The median notal wing process, differentiated into median and posteromedian notal wing processes, has the primary role in wing operation.

The Paraneoptera and Oligoneoptera are synapomorphic with regard to the presence of a true scutellum with a paired muscle *t13*.

The data presented above demonstrates that muscle *t13* was lost secondarily in the Paraneoptera. Moreover, this muscle is present in the Ephemeroptera and seems to have been present in the Meganeurida. However, the scutoscutellar suture branches are not closed up in mayflies, which sheds doubt on whether the presence of muscles connected with these branches can justifiably be regarded as evidence of the relatedness between the Ephemeroptera and Oligoneoptera. Hence only the Paraneoptera and Oligoneoptera may be reliably considered to be close to each other on the basis of the existence of a true scutellum. The Polyneoptera do not have a true scutellum; however, in these insects the pseudoscutellum is developed to focus the effort from contraction of the dorsal

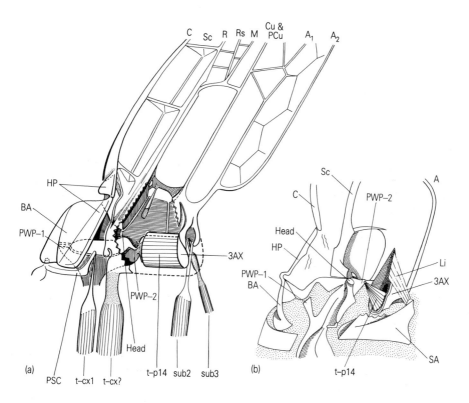

Fig. 10.6 Lateral view of the thorax of a dragonfly and a mayfly showing the arrangement of sclerites and muscle *t–p14* about the pleural wing process. (a) Dragonfly from the family Aeschnidae, inside view; sub2 and sub3 are the second and third subalar muscles (after Pfau 1986). (b) The mayfly *Ephemera vulgata*, outside view.

longitudinal muscles on to the anterior wing processes.

Another synapomorphic feature of the Para- and Oligoneoptera is the presence of the third axillary sclerite muscle *t–p13*. Formation of this muscle was evidently associated with improvement of the mechanism for folding the resting wings. Comparative analysis of tergal organization in the Psocoptera and Megaloptera demonstrates that devices for roof-like wing folding already existed at this level—the orientation of the oblique sutures, well-developed platforms lying distally from the oblique sutures (the future tergal wing grooves), and a shortened recurrent scutoscutellar suture. Further adaptations for roof-like wing folding proceeded independently in the two groups and had different end results. However, it should be emphasized that this adaptation began in common ancestors such as, perhaps, some of the Hypoperlida.

10.4.2 The axillary apparatus

In the axillary region of the insect wing, there are three zones which differ in their structural stability. The most conservative of these is the proximal zone which includes the first, second, and third axillary sclerites, the basisubcostale and basiradiale. The middle zone is more variable. It includes the radial and postcubital stems, the proximal medial plate, the basal region of the jugal fold, and the vertical hinge, which forms the top of a triangle which turns over when the wings are folded. The distal zone includes the sclerotized *Sc–R* bridge, the distal medial plate, and the medial and cubital vein bases. The torsional hinge is situated in this zone, and is associated with regular changes in the distal zone that depend on the increase in the ability of the wing to rotate around its longitudinal axis. This increase is primarily achieved by moving the posterior veins anteriorly. The most extreme expression of this tendency occurs in the Diptera and Hymenoptera; that is, in insects that have changed the way their wings are folded. One reason for this change may have been the need for greater freedom for pronational and supinational wing rotation and, consequently, a reduction in the relative size of the distal medial plate. Of

the groups with roof-like wing folding, the most perfect torsional hinge structure is found in the Neuroptera. The connection between the third axillary and the subalar sclerites was weakened in polyneopteran and oligoneopteran lineages, with all the consequent changes of the wing base structure.

The Plecoptera have the most primitive axillary zone structure of all extant insect groups. Their normal axillary apparatus differs from the generalized form in that it lacks the distal medial plate and the humeral fold is positioned differently. The uniqueness of the axillary apparatus structure in the Ephemeroptera and Odonata provides evidence for their early divergence and profound specialization, in which mastery of passive forms of flight played an important role. The prime influence in paraneopteran axillary apparatus evolution was the improvement of roof-like wing folding, whereas in the Oligoneoptera improvements in the flapping mechanism were equally important. It was in the Oligoneoptera that adaptations which provided automatic control of wingbeat parameters developed in the axillary zone of the wing.

10.4.3 Towards an auto-oscillating system

The increased efficiency of effort transmission from the muscles to the wings implies that mechanisms based on the use of elastic properties of the skeleton were formed and then improved. The three groups of prerequisites for these are as follows:

(1)   an increase in wingbeat frequency, implying the more frequent use of elastic deformation forces;

(2)   differentiation of the wing muscles into two groups: those producing thrust, which have a high contraction rate; and those controlling wingbeat parameters via changes in the positions of skeletal elements, which have a low contraction rate;

(3)   the formation of various elastic elements of the skeleton such as joint heads and highly resilient regions.

An important landmark in the evolution of wing movement automation was the fusing of separate parts of the pterothoracic skeleton with regions of the prothorax and abdomen into an integral rigid framework, and the appearance of functional and subsequently of anatomical two-wingedness.

In mayflies, a joint head was formed on the top of the mesothoracic pleural process (Fig. 10.6b). The complex shape of this head makes the wing move downwards and twist, and pronate automatically, without a direct muscular contraction. In the honey bee, the arrangement of sclerites in the wing base is controlled by special muscles; a change in this arrangement ensures automatic wing pronation at certain points of the wingbeat. The axillary lever plays an important part in the control of pronation; it is provided with muscles, and regulates the position of the scutellar arm relative to the first axillary sclerite and the pleural process. Active use of skeletal elastic forces in wing movement can be seen most strikingly in the 'radial stop' mechanism in the higher Diptera described by Miyan and Ewing (1985). The action of this mechanism includes the snap of the first axillary sclerite during the downstroke, and stopping the radial stem against the tip of the pleural process, which stretches dorsoventral muscles at the end of downstroke and thus enhances the effect of their contraction on lifting the wing. This means that energy is stored during the downstroke for release in the upstroke. Small tergopleural muscles (mostly those attached to the third and fourth axillary sclerites) control the wingbeat parameters of the right and left wings independently. In the higher Hymenoptera and Diptera, these muscles change the forces produced by the wing, and hence determine the flight path, by varying the stroke plane angle and the rate of wing pronation.

While studying the relationship between wingbeat frequency and atmospheric pressure, Chadwick (1951) and Sotavalta (1952) concluded that wingbeat frequency remains virtually unchanged, even with a sharp decrease in air density, which is especially pronounced in large insects with a high wingbeat frequency. Traumatic shortening of the wings or artificial wing loading in the Diptera and Hymenoptera led to a growth or a fall in wingbeat frequency respectively. That shortening of the wing changed the wingbeat frequency of both wings as well as the stability of the wingbeat frequency at various atmospheric pressures indicates the existence of a resonant frequency in the wing-body system.

In other words, the wing apparatus of higher insects may be regarded as a system connected by elastic links which works as an auto-oscillator. Such a system only consumes the energy necessary for compensation of friction losses. Work required for wing acceleration is performed at the expense of energy accumulated by the elastic links—the muscles and resilient regions of the skeleton. The kinetic energy accumulated by the wings when slowing down is transformed into potential energy for the resilient elements. The ability to store kinetic energy in skeletal elements distinguishes insects from birds, whose wing joints lack resilient structures. The insect exoskeleton limits increases in body size; however, it is the exoskeleton that enables elastic forces to be used in wing movement. Making use of these forces became a major determinant of the direction of the evolution of the wing apparatus. By rearranging the skeletal elements, the small tergopleural and pleurosternal muscles of the higher Diptera and Hymenoptera can change the tension of the thoracic walls and hence control wingbeat frequency.

10.5 The influence of body size on wing apparatus operation: two strategies in the evolution of the wing

Reduction of body size, a widespread phenomenon in the evolution of winged insects, significantly affects the structure and function of the wing apparatus. From palaeontological evidence, early winged insects tended to be larger than recent ones; however, early insects still ranged from very large (for example *Zdenekia*, whose forewing was about 65 mm long) to small (for example *Metrepator*, whose forewing was 7 mm or less). A significant increase in body size (gigantism) occurred later, and affected mostly the Palaeodictyoptera and Meganeurida. Body size gradually became reduced in the course of evolution in insects such as the Odonata, Ephemeroptera, and Blattodea. Some groups consisted of small insects from the very start of their palaeontological records—cicadas, dipterans, caddisflies, moths, and perhaps bugs. Consequently, there must have been subsequent increases in body size in these groups.

During the critical period of the formation of particular groups, body size must have been especially influential in the organization of the wing apparatus. This is exemplified by the modern Diptera, whose wings lack a true clavus and seem never to have had one, since supinational deformation of small wings does not involve supinational twisting. Although there are relatively large Diptera today their wings also lack a true clavus, reflecting the wing supination of their remote ancestors. Thus, at the dawn of a group's history very stable features of the wing apparatus are formed which often depend on the body size of the ancestral group members.

The effects of a reduction in body size also depend on the level of wing apparatus organization at which the process began. Thus, the body size decrease in the Permothemistida, from the gigantic ancestral palaeodictyopterans to medium-sized specimens (with a wing length of between 9 and 17 mm), led to a form of wing organization which is very close to that of extremely small Ephemeroptera, while mayfly wings, which are of the same size, are strongly pleated, triangular, and richly veined, with numerous transverse veins.

The clearest and most striking consequence of the decrease in absolute body size is the increase in wingbeat frequency, and though the reverse also occurs, as for example, in the case of stoneflies (Brodsky 1982), it is an exception rather than the rule. The relationship between wingbeat frequency and wing length in closely related insects can be described by a finite equation, such as $n = 822 \cdot l^{-0.725}$ (Belton and Costello 1980). The wingbeat frequency of the small dustywing *Coniopteryx pygmaea* (with a forewing length of 2.5 mm) can reach 80 Hz. Various small Diptera with wing lengths ranging from 2 to 7 mm have

wingbeat frequencies ranging from 150 to 600 Hz. In the tiny whitefly *Trialeurodes vaporariorum*, whose wing length is less than 1.5 mm, wingbeat frequencies ranging from 143 to 181 Hz have been recorded (Wootton and Newman 1979).

Despite the universal nature of this phenomenon its causes are not fully clear. Two main causes can be distinguished. In the first place, there is a decrease (due to a poorer ratio, at any angle of attack, between the coefficients of wing lift and drag) in the relative aerodynamic forces produced during a single wingbeat cycle. In the second place, there is a decrease in the inertia and the oscillation period of the wing, and an increase in capacity of the wing muscles to act fast by shortening the muscle contraction and relaxation periods. With few exceptions the increase in the capacity of the muscles to act fast does not affect their structure.

It is commonly held (Pringle 1957, and others) that an increase in oscillation frequency must be accompanied by an increase in wing movement velocity and therefore in the Reynolds numbers which characterize the movement. However small-sized species have higher angular velocities and lower oscillating velocities. Hence no increase in wingbeat frequency could compensate for the decrease in Reynolds number caused by their small body size. In the chalcid wasp *Encarsia formosa* (with a wing length of 0.6 mm), wingbeat frequency reaches 370 Hz, whereas the *Re* value is as little as 15 (Weis-Fogh 1975). At $Re = 200$ the wing lift to drag ratio is already 1:1, and it does not depend significantly on the orientation of the wing in the airflow. Moreover, vortex formation is impaired at low Reynolds numbers. According to Schlichting (1974), the Kármán vortex track is still formed at $Re = 65$, but at a distance, rather than immediately, behind the body. This circumstance forces small insects to use supplementary mechanisms for vortex production, such as the wing clap over the dorsum. It was more than chance that this mechanism was first described in a very small insect, a chalcid wasp. The clap at the top of the stroke is an essential attribute of flight in all small insects.

With a significant decrease in Reynolds number, the wing lift to drag ratio becomes less

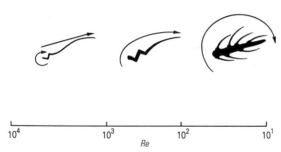

$$10^4 \qquad\qquad 10^3 \qquad\qquad 10^2 \qquad\qquad 10^1$$
Re

Fig. 10.7 Presumed flow patterns around the wing profile during the downstroke, at different Reynolds numbers.

than one. This fact has led people to suggest that small insects flying in the 'twilight zone' (*Re* < 100) use a special mechanism based on wing drag (Horridge 1956, Norberg 1972*a*, and others). Bennett (1973) suggested that flight could be achieved via a difference in velocities: during the downstroke velocity and hence wing drag are higher than during the upstroke. Until the flight aerodynamics of small insects has been studied, the normal range of Reynolds numbers for these insects will remain a 'twilight zone'. Taking into account indirect data such as the orientation of the elements of wing microrelief, wing-beat parameters, and deformation peculiarities, it may nevertheless be suggested that there is a certain regularity in the change in air currents around a wing as the *Re* value decreases (Fig. 10.7). Impaired vortex formation at low Reynolds numbers forces small insects to resort to energetic wing movements—hence the large wing beat amplitude, the indispensable wing clap at the top of the trajectory, and the use of right and left wings as part of an integral propulsive device whose influence on the environment leads to the formation of a common vortex ring around the body. Correspondingly, the importance of deformations in vortex formation diminishes in small insects, and the organization of wing microrelief is different.

The fact that inertial forces begin to play a less important role during flight results in simplification of the wing structure of small insects. Wing deformation during the wingbeat cycle is reduced, the longitudinal folds are less marked, and the tip deflection line moves towards the wing base, which can be clearly seen when comparing the wings of large and small stoneflies, flies, and many other insects. In the Trichoptera, Lepidoptera, and Hymenoptera, a fringe of hairs has developed along the wing border. Owing to the relatively high viscosity of air, this fringe functionally replaces the continuous membrane.

The influence of viscosity on the airflow around small insects is seen not only in the impairment of vortex formation but also in a drop in the lift needed to counterbalance the body weight. Since a small insect mainly has to produce thrust, it orientates its stroke plane more closely to the vertical compared with larger related forms. A vertical stroke plane results in the wings moving downwards at large angles of attack. The axis of centres of aerodynamic pressure undergoes a secondary shift towards the middle of the wing chord, which leads to the wing tip region becoming rounder, this phenomenon being more or less pronounced in different groups (Fig. 10.8). Wing costalization is retained. Having appeared earlier as a response to the need for strengthening the leading edge of the wing as it moved along a sloping trajectory at a small angle of attack, costalization now lost the role of mechanical support and is associated with ensuring the action of the complex multihinge supinational mechanism.

One further type of supinational deformation developed in parallel with the reduction in body size: a strong tip deflection, symmetrical about the longitudinal wing axis, along the transverse fold without interruption of the veins (only supporting veins can be interrupted)—a mechanism which does not appear in large insects. The shift of the transverse fold towards the base is accompanied by a reduction in the supporting elements, and leads to a symmetrical wing shape (Fig. 10.8e). When this type of deformation is strongly pronounced, the greater part of the wing is deflected downwards at the bottom of the stroke. This type of supination has appeared, with reduction in body size, in insects of various groups such as the Hymenoptera, Diptera, Trichoptera, and Lepidoptera, and begins to dominate over other

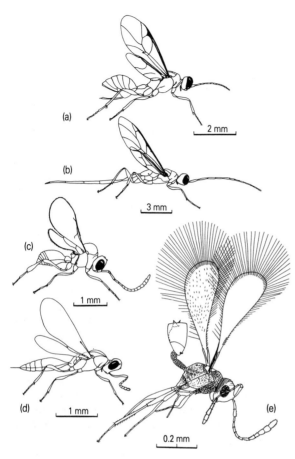

Fig. 10.8 Wing shape in various hymenopterans (after Rasnitsyn 1980*a*). Decreases in body dimensions direct the vector of evolution towards a sculling strategy. (a) *Austronia* sp., (b) *Monomachus* sp., (c) *Ismarus* sp., (d) *Brachinostemma* sp., (e) *Palaeomymar* sp.

types of supinational deformation at a definite level of body size reduction in each of the orders.

In very small insects such as thrips, some beetles, and hymenopterans a pinnate wing is typical—a strap with one or two supporting veins and a long fringe of hairs along its edges. Some tiny hymenopterans can even swim with these wings. Although swimming has not yet been studied in detail, its wing kinematics seems

not to differ significantly from that during flight: moving down, the wings produce a stroke; moving up, either they bend or their hairs are bent. Such flight is rather energy-consuming; however, it is this mechanism that allows tiny insects to hover, to fly slowly from site to site, and to perform long-distance migrations in air currents.

Change in wing shape and structure with reduction in insect body size is a universal phenomenon that occurs in various groups, and that may be observed even when body size reduction is not as marked as in the insects discussed above. Danforth (1989), after having analysed the influence of body size on wing shape and structure in the Hymenoptera, came to the conclusion that as insect body size (body weight) increases, the ratio of pterostigma area to wing area decreases, the wing cells shift towards the wing tip, the wing becomes more elongate, the maximum wing width shifts towards the wing base, and wing loading increases. All these changes can be explained by the fact that as body size decreases the insect wing begins to operate in a 'scull regime'—the propulsive force is created by drag but not by lift. Thus, in the evolution of insect wings two extreme states, two different strategies exist: the scull and the wing. Body size decrease affects wing evolution so that the wing transforms into a scull: the aspect ratio is reduced, the wing tip becomes rounded, and the centre of mass shifts closer to the wing tip. Due to changes in its deformational character, the wing cells shift towards the basal end and the relative size of the pterostigma increases.

It is interesting to note that evolution of the wings according to one of the two strategies is not necessarily correlated with body size. According to observations by the hymenopterologist Tobias (personal communication), in woodland species of Ichneumonidae which live in conditions of still air and high humidity the scull strategy predominates, whereas in species which inhabit open arid spaces the wing strategy is prevalent. Thus high air humidity and the absence of wind appear to simulate a decrease in body size.

10.6 Vortex formation, and particular features of insect flight

The ancestors of winged insects moved in the air by means of proto-wings which performed progressive oscillations in a plane perpendicular to the oncoming steady airflow. More affective thrust generation requires a combination of progressive and rotational oscillations; the latter were brought about by the influence of wing inertial forces at the extreme points of a wingbeat cycle, and allowed insects to fly independently of oncoming wind currents. A further improvement in flight efficiency was achieved by elongating the wings and increasing the wingbeat amplitude.

During insect flight the flapping wings leave behind them a wake consisting of vortex rings. Unlike in birds, the shape of the vortex wake shows significant variations in insects. A vortex wake consisting of discrete vortex rings was recorded for broad-winged butterflies (Ellington 1980, in the migratory flight of the peacock butterfly). Vortex rings have also been described in the wake of small passerines (Kokshaysky 1979) and for a pigeon and a jackdaw in slow flight (Spedding *et al.* 1984, Spedding 1986). These animals have much in common: low wing aspect ratios, low wingbeat frequencies, and large wingbeat amplitude. The advance ratio (the distance covered per flap) is small or equal to zero in all these cases. This vortex wake structure can reasonably be explained by wing shape and the stroke parameters, although the flying passerine produces vortex rings only during the downstroke, while the butterfly creates them both during the downstroke and the upstroke. An increase in wingbeat frequency causes changes in the structure of the wake. At first, only two vortex rings are coupled, as in the feeding flight of the peacock butterfly; then the flapping wings shed a chain of coupled vortex rings, as in the tethered flight of the Essex skipper; and finally, as in the crane fly, the wings produce a pair of chains consisting of small uncoupled vortex rings. The evolutionary development of the flight of winged insects began at the stage when the wake behind the flying insect was represented by a chain of coupled vortex rings (Fig. 10.9a).

The significance of the circulatory airflow produced during wingbeats varies depending on how the aerodynamic forces are generated by the various types of wing interaction with the oncoming air. The significance of this bound circulation is relatively great in insects with low wingbeat frequencies, but much less so in species with high wingbeat frequencies. Considering the circulation of the starting vortex (given that it is equal in magnitude but opposite in direction to the circulation around the wing), the circulation around the wing is rather high in insects with low wingbeat frequencies. During the downstroke, the bound circulation produces a flow on the dorsal wing surface. In insects with low wingbeat frequencies, this flow covers a significant part of the wing, while in insects with high wingbeat frequencies it consists only of a relatively narrow twisting stream which crosses the wing at its broadest part. To obtain a clearer idea of the role of the circulation in the creation of aerodynamic forces in insects with high wingbeat frequencies, it is necessary to resort to quantitative data about the velocity of airflows close to the wing surface and in its wake.

In spite of the relatively important role of bound circulation in the generation of aerodynamic forces in insects with low wingbeat frequencies, their flight can nevertheless not be explained from the viewpoint of classical steady-state aerodynamics. There are three objections to such an explanation.

The first objection is the complex nature of the circulation that is generated by a flapping wing: first, because the angle at which the free airflow meets the wing varies throughout the half-cycle and is sometimes large, and second, because flapping movement produces centrifugal forces that tend to move the air outwards down the wing. The most complete description a quasi-steady approach can provide is for the flight of some Orthoptera whose hindwings move at small angles of attack and whose polar diagrams are 'pointed' (see Fig. 3.3b, p.53). The circulation around the wings of these insects seems to be fuller and steadier relative to other insects with similar low wingbeat frequencies. It was not an accident that the best match between calculated data and measured forces was found in a desert locust (Jensen 1956).

The second objection is that the steady-state aerodynamic approach is hindered by the influence of vortices on the velocity and direction of the airflow around the wings. An important feature of flight aerodynamics in insects with low wingbeat frequencies is that the dorsal vortex, unlike the starting vortex of the supporting plane, is not carried away by the flow but rather is held in close to the wings, influencing their airflows during the entire half-cycle of a wingbeat. This statement is also true for the ventral vortex.

The third objection is that the airflow acceleration caused by a vortex ring formed during the preceding half-cycle leads to compensation for the negative lift which corresponds with lifting the wing.

Further aerodynamic rules come into force in the flight of insects with high wingbeat frequencies: the velocity of bound circulation is determined mainly by the tip vortex, which also creates a small vortex ring which is cast off by the end of the half-cycle. Here it is possible to talk about a new unsteady mechanism for the generation of forces: the predomination of the tip vortex in the airflow around the wing is caused by unsteadiness of the wing motion itself (i.e. by the velocity vector turning the opposite direction at the extreme points of a wingbeat). In contrast, bound circulation is a typical attribute of steady motion during which the wing neither accelerates nor is inhibited. Even in this, unsteady effects are widely employed for generation of aerodynamic forces. It takes some time (or distance) to build up the circulation around the wing, so if the wing moves abruptly from its resting position, lift does not peak until the wing has travelled about one transverse length of the aerofoil (one wing-chord length). To create the circulation immediately after the onset of the downstroke, an insect has to pronate its wings thus producing the leading edge vortex which closes on the starting vortex. This brings us back

to the idea that pronation is the most critical stage of the wing movement which determines the fate of vortex rings shed from the wings.

In the course of evolution there is a steady increase in the importance of unsteady effects in the creation of aerodynamic forces. Thus, when a straight vortex street is formed behind the

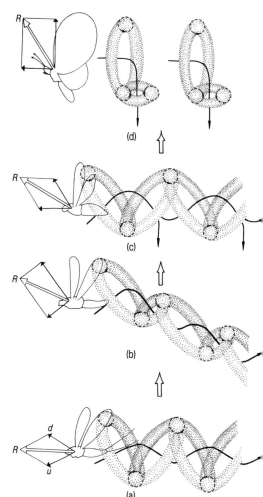

Fig. 10.9 Evolution of the vortex wake pattern in insects. The open arrow (R) corresponds to the force acting on the insect during forward flapping flight. R is the reaction to the vortex ring momentum of the downstroke (d) and upstroke (u). The bold arrow indicates the direction of wake central flow. For explanation of a–d see text.

body during flight (Fig. 10.9a), the generation of forces both during the downstroke and during the upstroke can be explained by a quasi-steady aerofoil action. During formation of a slanted vortex street (Fig. 10.9b) in functionally four-winged insects the situation does not change fundamentally, although here the force impulse created during the downstroke exceeds that generated during the upstroke.

When a false straight vortex street is formed behind an insect (Fig. 10.9c—imagos of mayflies, functionally two-winged insects, and stoneflies), the generation of forces during the upstroke can be explained by an unsteady aerofoil action: during the upstroke a flow formed by the downstroke ring dominates the airflow around the wings. Moreover, during pronation the insect wings push the ring chain backwards, as a result of which the insect is pushed forward. Consequently, formation of forces at the upper point of the stroke can be explained using the principles of a reactive mechanism. The role of this mechanism becomes more significant in broad-winged butterflies since they throw off discrete vortex rings. In these insects, a chain of vortex rings first becomes disconnected at the upper point of the stroke (a feeding flight, Fig. 10.9d) and then again, due to the strong clap of the wings, at the lower point of the stroke (a migration flight). Finally, in insects with high wingbeat frequencies, casting off vortex rings is a predominant mechanism in the generation of forces.

When considering the generation of forces in flapping flight we must take account of the mechanism of lift production in the second type of hovering. Because of the low stroke amplitude used, the second type of hovering has been a stumbling-block on which several theories of flapping flight have collapsed. Weis-Fogh (1973) made the first attempt to explain how forces are generated by hovering Syrphidae. This author believed that the wings of a hover fly move both down- and upwards along an inclined trajectory with positive values of angle of attack so that the forces normally generated are mutually eliminated (the downstroke force vector points upwards and forwards, while during the upstroke it points downwards and backwards).

As the wings rotate around their longitudinal axes at the extreme points of their trajectory, vortices are formed that generate a force impulse pointing upwards. It is this force, named the 'flip-lift' by Weis-Fogh, that counterbalances the body weight of the hover fly when hovering. Ellington (1984), having demonstrated that in fact the insect wings are lifted at a negative angle of attack, suggested that a different mechanism underlies the second type of hovering. This mechanism is based on rotational effects, when during rotation of the wing around its longitudinal axis at the upper point of the stroke, the flow separates from the costal edge of the wing, thus leading to vortex formation on the upper surface of the wing. This vortex enhances the circulation around the wing markedly, which explains the unusually high lift coefficient value.

The data obtained using flow visualization around flying insects (Brodsky 1986) demonstrated that flow separation, with consequent vortex formation on the upper surface of the wing, did not take place. In insects with relatively high wingbeat frequencies the vortex on the upper surface of the wing does indeed form, but it is associated with tip effects and is localized close to the wing tip.

Taking account of specific features of insect wing kinematics and deformation during the second type of hovering, as well as of the possibility that small vortex rings may separate from the wing on one side, we shall try to present an integrated picture of the generation of forces in the second type of hovering.

When, in the process of evolution, the wing-beat amplitude became smaller and wing rotation around the longitudinal axis at the extremes of the trajectory was improved, it became possible to manage the casting off of the vortex rings which were formed at each half-cycle. The system of coupled vortex rings has to be shed at right angles to the stroke plane (Fig. 10.10a), whereas in the third type of hovering the many isolated rings are cast off parallel to the stroke plane (Fig. 10.10c). In the Syrphidae and in most Diptera Orthorrhapha, the wings undergo supinational rotation at the lower point of the trajectory (that is, a very rapid wave along the rear margin) so that the wing plane rotates

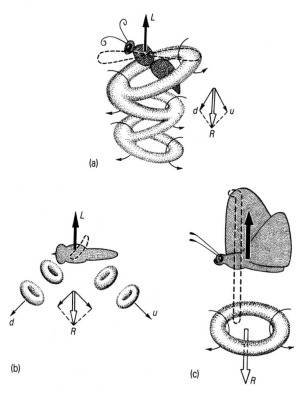

(a)

(b)

(c)

Fig. 10.10 Vortex wake pattern in different types of hovering. (a), (b), and (c) are the first, second, and third types of hovering, respectively. The open arrow (R) is the resultant force from the downstroke (d) and upstroke (u) vortex ring momentum whose reaction generates lift, which counterbalances the body weight. The path of the forewing is shown by a broken line.

rapidly around the leading edge. As a result of this rotation, the small downstroke vortex rings are cast off in the same direction as the stroke plane (Fig. 10.10b). At the end of the upstroke the wings of a hover fly undergo pronational bending; the anterior functional field rotates around the posterior one, and, as a result, the rings are cast off at right angles to the stroke plane (Fig. 10.10c). Thus, four vortex streets are formed beneath a hovering hover fly, two on each side of the body. The reaction to the

momentum of these vortex rings, cast in differe directions, generates lift which counterbaland the hover fly's body weight during the seco type of hovering. Forces are generated sin larly in anisopterous Odonata in the seco type of hovering, with each of the four wir functioning as an independent vortex generat The deformational wave which propagates alo the wings at supination plays an important r in the formation of small diameter vortex ring In the anisopterous odonate hovering in t regime described above (see Fig. 6.9, p.11 the final picture of vortex formation is ide tical to that of hovering hover flies, althou the actual mechanisms of vortex formation a different in the two cases. In the Syrphida small vortex rings are shed off the wings both the uppermost and lowermost points the stroke, whereas in the Anisoptera they lea the wings only at the lowermost point. In oth words, hover flies, using only one pair of wing achieve the same effect as Anisoptera which hav two pairs of independently operating wings.

The appearance of a new mechanism for ge erating and managing the shedding of vort rings radically expanded the flying abilities insects. There was a dramatic increase in flig velocity and manoeuvrability. Many represent tives of historically young orders are able hover, thus affecting both specific features an patterns of flight. The behaviour of the particula insect species exerts an especially strong influ ence on its general flight pattern. Thus durin searching flight a crane fly appears to fly wit yawing movements but for other ethologic functions this insect may hover and, as it gain altitude, does not turn its head up as does th green lacewing. Higher Diptera are able to fl sideways but, if necessary, they undertake quick manoeuvre to align the longitudinal bod axis with their course of flight.

According to Hertel (1966), ideal flight in cludes vertical take-off and vertical landing vertical climb and descent, hovering withou loss of height, and movements backwards an forwards from the stationary position. All thes features are highly typical of insect flight, an furthermore are performed at whatever speed i necessary. Ideal flight also requires the cruisin speed to be as high as possible. Hence in the course of evolution, insects have achieved what has been deemed ideal flight; they are able to fly backwards at low speed, to fly at zero speed, and to fly forwards as fast as possible.

In addition, insects can turn their longitudinal axis of the body abruptly towards the direction of flight, they can climb without turning the body upwards, and finally can hover at a very small stroke angle.

10.7 Conclusion

The insect wing apparatus is a complex biomechanical system that allows movement in air. In the course of evolution, new structures have appeared and old ones have become simplified. The earlier a structure appeared in the evolution of the wing apparatus, the more similarly it is organized in all winged insects. Thus, the horizontal hinge of the wing root is more structurally uniform than the torsional hinge. The creases that allow the rear part of the wing to fold run similarly in all insects which are able to fold their wings, whereas the vannal folds that appeared later are highly diverse. The most ancient of the deformable folds, the cubital fold, is noted for its stability of position in the wing root, and determines the uniformly concave state of CuP.

In general, structural changes followed on from and were influenced by functional changes. For example, the folds and grooves formed during wing strokes are uninterrupted; the necessary conditions for their formation were created by changes in the position of the veins, their fusion, and development of additional veins. Similarly, the synchrony of operation of the fore- and hindwing pairs was originally made possible by a common rhythm of muscle contraction, and was only later underpinned by special coupling mechanisms.

Evolution proceeds in a helix. During this process the characteristic features of ancestral forms may be repeated in their descendants, but on a new basis. Hence the synchronous flapping in the first winged insects was replaced by phase shifted movement of the wing pairs. Later, after coupling devices had appeared, synchronous movement of the wing pairs became more complete. Similarly, stroke amplitude first increased and then decreased during the course of evolution. The development of torsional hinges and the consequent increase in the freedom of rotation of the wings about their longitudinal axes was a prerequisite for the stroke amplitude decrease. The weakening of the posterior notal wing process in the Oligoneoptera meant that the connection between the third axillary sclerite and the notum was lost; however, in the Hymenoptera Aculeata and Diptera Brachycera the posterior notal wing process, after fusing with the lateropostnotum, became an active participant in wing operation all over again.

Decreases in body size, for whatever reason, are associated with an increase in stroke amplitude and wingbeat frequency, simplification of wing structure, changes in the orientation of the elements of the microstructure, and the appearance of a new type of supinational deformation, namely symmetrical downward deflection of the wing tip. Reduced body size resulted in some return to ancestral features: upward flight with the body vertical, increased stroke amplitude, and the adoption of a trajectory perpendicular to the longitudinal body axis by the wings.

Postscript

The evolutionary origin of insect flight must have involved flapping flight. This is the only reasonable explanation for the increase in wing length between wingless ancestors and the first winged insects. Flapping flight meant that insects had to face a number of problems, whose solutions resulted in the development of the existing wide variety of types of wing apparatus.

The first, and historically earliest problem was to overcome the inertial forces which developed at the end of each half-cycle. The struggle against inertial forces gave rise to supinational deformation, first as gradual supination, then as supinational twisting, then as supinational bending, and finally wave-like deformation. The development of each of these stages caused a transformation in wing structure: reduction of the dense archedictyon network and the many transverse veins, fusion of the longitudinal veins, and isolation of functional wing regions so that many of the wing surface deformations, whether from tip to base or vice versa, became automatic. The main result of the changes in wing structure was the formation of a structure well adapted for pronational deformation, whose most advanced form is the rear marginal wave which passes from the tip to the base of the wing thereby bringing the entire wing plate into the pronated state. Transformation of the character of pronation, in turn, enabled the nature of the interaction between the wing and the airflow to change, since the fate of vortex rings shed from the wings at the end of the half-cycle depends on the specific character of the pronation. The most critical of these aerodynamic changes was the closing of the vortex rings around the wing (or the coupled wing pair) on one side of the body, which is directly associated with a sharp increase in manoeuvrability.

The second problem faced by insects during the first steps in mastering the airspace was the necessity to reduce the interference between the fore- and hindwing pairs. One of the fir solutions to this problem was the slight shor ening of the hindwings and the appearance of difference in stroke amplitude between fore- an hindwings. From then on the transformatior led towards functionally four-winged flight, i which a common system of vortex rings formed behind both wing pairs, which mov out of phase. Later on functionally four-winge flight was replaced by functionally two-winge flight, which occurred at first by means of common rhythm of wing muscle contraction an subsequently by special coupling mechanism: Coupling of the wing pairs was in turn assoc ated with an increase in wingbeat frequency an a consequent reduction in vortex ring size.

Successful mastery of the airspace was mainl a function of the solution of the third prob lem, namely the development of energy-savin modes of flight, which could be achieved b the ancient insects, from an aerodynamic view point, in the following two ways: 1) by increas ing the stroke amplitude and 2) by lengthen ing the wings. Both ways appeared promising but resulted in a different final wing appa ratus design.

Increased stroke amplitude led to the devel opment of a wing apparatus with the following features:

(1) relatively short wings which perform high frequency and low-amplitude strokes bu with a large angle of rotation around the longitudinal wing axis;
(2) torsional hinges with a complex structure ir the distal zone of the axillary apparatus;
(3) anatomical two-wingedness;
(4) the main role in the downstroke played by the indirect flight muscles;
(5) active use of the elastic forces of the skeleton.
(6) relatively weak supinational wing deforma tion and, in contrast, strong pronationa

deformation, expressed by a wave passing over the wing surface;

(7) high velocity and manoeuvrability of flight, using hovering and cascades of turns, each change in direction being accompanied by immediate turning of the longitudinal body axis.

All these features occur in the wing apparatus of the Diptera Cyclorrhapha.

The alternative route to mastery of the air led to the following wing apparatus organization:

(1) relatively long wings performing low-frequency and low-amplitude strokes;

(2) independent movement of both fore- and hindwing pairs and of the wings on the right and left sides;

(3) wings connected with the thorax rigidly and wing rotation along the longitudinal axis made possible by several hinges of relatively late origin, situated distally in the wing plate;

(4) the downstroke performed by means of contraction of the direct flight muscles;

(5) high velocity, manoeuvrable flight with frequent use of hovering and gliding, and rapid changes in flight direction either accompanied by sudden stops (as in the anisopterous Odonata), or performed at any angle to the longitudinal body axis (as in the zygopterous Odonata).

Thus, having considered the patterns and forms of insect flight, wing apparatus structure, the character of wing movement, and the interaction of the wings with the air, we attempted to answer the question posed by Snodgrass more than a half of century ago: 'How do the insects fly?' We have done our best to give the reader a complete picture not only of insect flight in general but also of how diversified this flight can be, what the differences in flight mechanisms are among the various groups, what principles of force generation are required for movement in air, and the changes which the flight-providing system underwent during its evolution. We have obtained answers to many of these questions but many new ones have arisen, and the directions in which our efforts should be concentrated to

investigate this complex and multilevel phenomenon further have become more clearly focused.

The study of specific features of insect flight, as that of other biological aeromechanical objects, may be regarded from two viewpoints. First of all, a correct interpretation of the phenomena which take place during the movement of insects in air may provide valuable and useful data for the solution of purely biological problems such as those about insect morphology, the interactions between the various parts of the insect, and the relationships between the insect's structure and functioning. Analysis of the functional roles of isolated elements of the wing apparatus in the generation of forces is especially important for the solution of problems in insect evolution.

By studying the bioaerodynamics of insect flight, one can begin to understand the principal features and physical rules of flapping flight within a limited range of movement regimes. However, when one considers that insect flight, due to small body size and low velocity of movement, proceeds at low Reynolds numbers and that there is therefore too large a difference in scale between insects and flying man-made objects, it would be a great deal to expect that rules could be extrapolated to any field of technology. Still, from the viewpoint of engineering and a technological approach to the study of flight, the shape of the insect body and the airflow conditions, and arrangement of body parts and their construction are worth attention. Finally, it should be noted that most investigators engaged in insect bioaeromechanics are clearly attracted by its non-steady state phenomena, since the solution of questions about the interaction between the flapping wing and the airflow can only be found by this approach. Attempts to solve the problems of the bioaeromechanics of flapping flight based on steady-state aerodynamics may lead to a misapprehension of the structure and function of the wing apparatus of living flying objects, and particularly of insects.

Some of the characteristics of the non-steady nature of insect flight are, on the one hand, a source of incredible difficulty for experimentation and, on the other hand, of basic importance for flight as a physical phenomenon. The role of unsteady effects in insect flight and the evolution of the wing apparatus was emphasized by Weis-Fogh (1975): 'Application of non-steady principles makes possible innumerable modifications of flight mechanisms and wing shapes and one must take this into account when discussing the function and evolution of winged insects'. This quotation could serve as a testament to this outstanding investigator of insect flight, who died so prematurely.

The practical importance of insect flight research lies first of all in the application of knowledge of the principles and determinants of the interaction between the flapping wing and the airflow for designing and creating various kinds of non-steady propulsive devices. The main 'technical' achievement of winged insects is having overcome the inertial forces formed during wing strokes. Insects have 'learnt' not only how to overcome these forces but also how to use them to move the wings in the next stroke half-cycle. Humankind has expended much energy and many lives and achieved nothing but disappointment in its attempts to imitate flapping flight. The principal obstacles to the use of strokes for the generation of propulsive forces are the mighty inertial forces which become particularly high with the increased size of artificial flying machines. We have had to avoid the issue by using the principles of continuous rotation (the propeller) and high temperature fuel consumption (the reactive principle) to generate propulsive forces.

At present, it is hard to envisage how the many principles which have been learnt from insect flight aerodynamics could be used in flight technology; however, there are some aspects for which the solution of the problems may not be too distant. First, we may consider the shape and construction of insect wings, whose parallel grooves allow the air to move directly down the wing. There are also certain hidden advantages in the use of a wing whose geometry is variable. Dragonfly wings can create significant thrust with high efficiency, and a corresponding mechanism allows the hindwing to use the energy from the wake of the forewing. In addition, the study of a regime of intermittent

movement which consists of alternating wing strokes and long periods of gliding could play a critical role in the choice of an optimal functioning regime for man-made flight apparatus. The principle of recuperative energy exchange between the two wing pairs which is found in Odonata, has not yet been implemented in technical devices. If we are ever to create a flying machine with flapping wings, technical development of the strokes will have to be on a combination of proper and precessional rotations of the wing (see Chapter 2).

Relatively recently a well-known mechanical engineer Maxworthy (1981) identified some of the most pressing tasks in insect flight research:

(1) the analysis of flight aerodynamics at very low Reynolds numbers, for example in thrips or other feather-winged insects;

(2) an examination of particular features of the interaction between wing pairs when they move out of phase;

(3) the analysis of the role played by the flexibility (i.e. deformability) of the wings in the modification of the forces generated during wing movement;

(4) a study of the form of the wake of a flying insect;

(5) calculation of the energy and power required for flight.

The list of pressing tasks is clearly far from complete since only the aerodynamic aspects of flight have been taken into account. We have paid most attention to the second, third, and fourth problems. However, there still remain a vast amount to be done, even in these areas. First, there is a lack of detailed quantitative information on the distribution of velocities around the wings and behind the flying insect and there is an urgent need for data on the organization of the aerodynamic wake of the many insects with high wingbeat frequencies. The character of the air flow close to the surface of the flapping wing, the role of the microtexture in the distribution of air currents, and many other questions are all of great interest. Nowhere near all insect groups have been investigated. We know few details of the flight of such insects as the zygopterous Odonata or beetles.

Many interesting discoveries await anyone ready to investigate flight in these or other insect groups. The author hopes that the information presented in this book will be useful for all those who, fearless of the difficulties, decide to devote themselves to research into one of the most complex and fascinating phenomena in nature—insect flight.

References

Adams, Ph.A. (1958). The relationship of the Protoperlaria and the Endopterygota. *Psyche,* **65(4)**, 115–27.

Alexander, D.E. (1984). Unusual phase relationships between the forewings and hindwings in flying dragonflies. *J. exp. Biol.,* **109**, 379–83.

Anderson, M., Miller, T.A., and Schouest, L.P. (1985). The structure and physiology of the tergotrochanteral depressor muscle in the house fly. In *Insect locomotion*. Proceedings of Symposium 4.5, 17th International Congress of Entomology (ed. M. Gewecke and G. Wendler), pp. 97–102. Paul Parey, Berlin, Hamburg.

Antonova, O.A. and Brodsky, A.K. (1977). Functional pattern of wing muscles operation in the flight of the cockroach *Periplaneta americana. Zh. evol. bioch. physiol.,* **13**, 644–6. (In Russian.)

Antonova, O.A., Brodsky, A.K., and Ivanov, V.D. (1981). Wing kinematics in five insect species. *Zool. Zh.,* **6**, 506–19. (In Russian.)

Azuma, A. and Watanabe, T. (1988). Flight performance of a dragonfly. *J. exp. Biol.,* **137**, 221–52.

Azuma A., Azuma, S., Watanabe, T., and Furuta, T. (1985). Flight mechanics of a dragonfly. *J. exp. Biol.,* **116**, 79–107.

Badonnel, A. (1934). Recherche sur l'anatomie des Psoques. *Bull. Biol. Fr. Belg. Suppl.* **18**, 1–241.

Barber, S.R. and Pringle, J.W.S. (1966). Functional aspects of flight in belostomatid bugs. *Proc. R. Soc. (Lond.) (B),* **164**, 21–39.

Bauer, C.K. and Gewecke, M. (1985). Flight behaviour of the water beetle *Dytiscus marginalis* L. In *Insect locomotion*. Proceedings of Symposium 4.5, 17th International Congress of Entomology, (ed. M. Gewecke and G.Wendler), pp. 205–14. Paul Parey, Berlin, Hamburg.

Belton, P. and Costello, R.A. (1980). Flight sounds of some mosquitoes of Western Canada. *Entomol. Exp. Appl.,* **26(1)**, 105–14.

Benedetto, L.A. (1972). Plecopterenwanderungen in ufernahe Waldbereiche. *Ent. Tidskr.,* **93(4)**, 220–3.

Bennett, L. (1973). Effectiveness and flight of small insects. *Ann. Ent. Soc. Amer.,* **66(6)**, 1187–90.

Betts, C.R. (1986a). The comparative morphology of wings and axillae of selected Heteroptera. *J. Zool. Lond. (B),* **1**, 255–82.

Betts, C.R.(1986b). Functioning of the wings and axillary sclerites of Heteroptera during flight. *J. Zool. Lond.(B),* **1**, 283–301.

Bocharova-Messner, O.M. (1968). Peculiarities in the construction of the skeleto-muscular system of the pterothorax of the black cockroach. In *Functional morphology and embryology in insects,* (ed. D.M. Fedotov), pp.44–7. Nauka, Moscow. (In Russian.)

Bocharova-Messner, O.M. (1979). Wings of insects as flight organs. *Lecture series in memory of N.A. Kholodkovsky,* **30**, 3–40. (In Russian.)

Bocharova-Messner, O.M. (1982). Flapping surface of insect as an example of life-providing morphological model. *Problems of animal morphology development,* (ed. E.I. Vorobjova), pp. 128–39. Nauka, Moscow. (In Russian.)

Bocharova-Messner, O.M. and Aksyuk, T.S. (1981). Formation of a tunnel by wings of diurnal butterflies in flight (Lepidoptera, Rhopalocera). *Doklady AN SSSR,* **260(6)**, 1490–3. (In Russian.)

Boettiger, E.G. and Furshpan, E. (1952). The mechanics of flight movements in Diptera. *Biol. Bull. mar. biol. lab.,* **102**, 200–11.

Bramwell, A.R.S. (1976). *Helicopter dynamics*. Edward Arnold, London.

Brodsky, A.K. (1970a) On the role of wing pleating in insects. *Zh. evol. bioch. physiol.,* **6**, 470–1. (In Russian.)

Brodsky, A.K. (1970b). Organisation of the flight system in the mayfly *Ephemera vulgata* L. (Ephemeroptera). *Ent. obozr.,* **49(2)**, 307–15. (In Russian.)

Broadsky, A.K. (1971). Experimental study of flight in the mayfly *Ephemera vulgata* L. (Ephemeroptera). *Ent. obozr.,* **50(1)**, 43–50. (In Russian.)

Brodsky, A.K. (1974). Evolution of the flight

apparatus in mayflies (Ephemeroptera). *Ent. obozr.*, **53(2)**, 291–303. (In Russian.)

Brodsky, A.K. (1975). Wing kinematics of mayflies and an analysis of their power regulating mechanism. *Zool. Zh.*, **65(2)** 209–20. (in Russian.)

Brodsky, A.K. (1979*a*). Evolution of the wing apparatus in stoneflies (Plecoptera). Part II. Functional morphology of the axillary apparatus, skeleton and musculature. *Ent. obozr.*, **58(4)**, 705–15. (In Russian.)

Brodsky, A.K. (1979*b*). The origin and early stages in the evolution of the insect wing apparatus. *Lecture series in memory of N.A. Kholodkovsky*, **30**, 41–78. (In Russian.)

Brodsky, A.K. (1981) Aerodynamic peculiarities of insect flight. IV. Data on the flight of the mayfly *Ephemera vulgata* L. *Vestnik Leningr. Univ.* (Biology), **15**, 12–18. (In Russian.)

Brodsky, A.K. (1982). Evolution of the wing apparatus in stoneflies (Plecoptera). Part IV. Wing kinematics and general conclusion. *Ent. obozr.*, **61(3)**, 491–500. (In Russian.)

Brodsky, A.K. (1985). Kinematics of insect wing movement in rectilinear forward flight (a comparative study). *Ent. obozr.*, **64(1)**, 33–50. (In Russian.)

Brodsky, A.K. (1986). Flight of insects with high wingbeat frequencies. *Ent. obozr.*, **65(2)**, 269–79. (In Russian.)

Brodsky, A.K. (1987). Structure and function of the veins and folds in insect wing. *Trudy VEO*, **69**, 4–19. (In Russian.)

Brodsky, A.K. (1988). *Mechanics of insect flight and evolution of their wing apparatus.* Leningrad University Press, Leningrad (In Russian.)

Brodsky, A.K. (1989). Structure, function, and evolution of wing articulation in insects. *Lecture series in memory of N.A. Kholodkovsky*, **41**, 1–47 (In Russian.)

Brodsky, A.K. (1991). Vortex formation in the tethered flight of the peacock butterfly *Inachis io* L. (Lepidoptera, Nymphalidae) and some aspects of insect flight evolution. *J. exp. Biol.*, **161**, 77–95.

Brodsky, A.K. and Ivanov, V.P. (1973). Aerodynamic peculiarities of the flight of insects. I. Dependence of drag coefficients on Reynolds numbers. *Vestnik Leningr. Univ.* (Biology), **15**, 17–20. (In Russian.)

Brodsky, A.K. and Ivanov, V.P. (1974). Aerodynamic peculiarities of the flight of insects. II. Smoke spectra. *Vestnik Leningr. Univ.* (Biology), **3**, 16–21. (In Russian.)

Brodsky, A.K. and Ivanov, V.P. (1975). Aerodynamic peculiarities of the flight of insects. III. Flow around the wings of mayfly *Ephemera vulgata* L. (Ephemeroptera). *Vestnik Leningr. Univ.* (Biology), **3**, 7–10. (In Russian.)

Brodsky, A.K. and Ivanov, V.D. (1984). The role of vortices in insect flight. *Zool. Zh.*, **63**, 197–208. (In Russian.)

Brodsky, A.K. and Ivanov, V.D. (1986). Action of the axillary apparatus of the caddisfly. *Vestnik Zool.*, **4**, 68–74. (In Russian.)

Brodsky, A.K. and Vorobjov, N.N. (1990). Gliding of butterflies and role of the wing scale cover in their flight. *Ent. obozr.*, **69** (2), 241–56. (In Russian.)

Brodsky, A.K. and Judovsky, I.D. (in press). The recuperative principle in the action of dragonfly wing apparatus. *Doklady AN SSSR.*

Buckholz, R.H. (1986). Functional role of folds in wings of living systems. *Works of Amer. Soc. Engin.*, **1**, 279–86. (In Russian.)

Caple, G., Balda, R.P., and Williams, W.R. (1983). The physics of leaping animals and the evolution of preflight. *Amer. Natur.*, **121(4)**, 455–76.

Chadwick, L.E. (1940). The wing motion of the dragonfly. *Bull. Brooklyn Ent.Soc.*, **35**, 109–12.

Chadwick, L.E. (1951). Stroke amplitude as a function of air density in the flight of *Drosophila*. *Biol. Bull.*, **100**, 15–27.

Chapman, M.A. (1918). The basal connections of the tracheae of the wings of insects. *The wings of insects* (ed. J.H. Comstock), pp. 27–51. Comstock, New York.

Cloupeau, M., Devillers, J.F., and Devezeaux, D. (1979). Direct measurements of instantaneous lift in desert locust; comparison with Jensen's experiments on detached wings. *J. exp. Biol.*, **80**, 1–15.

Collett, T.S. and Land, M.F. (1975). Visual control of flight behaviour in the hoverfly, *Syritta pipiens* L. *J. Comp. Physiol.*, **99**, 1–66.

Comstock, J.H. (1918). *The wings of insects*. Comstock, New York.

Cullen, M.J. (1974). The distribution of asynchronous muscle in insects with particular reference to the Hemiptera: an electron microscope study. *J. Ent.*, **A49(1)**, 17–41.

Dalton, S. (1975) *Borne on the wind.* E.P. Dutton, New York.

Danforth, B.N. (1989). The evolution of hymenopteran wings: the importance of size. *J. Zool. Soc. Lond.*, **218**, 247–76.

David, C.T. (1978). The relationship between body angle and flight speed in free-flying *Drosophila*. *Physiol. Entomol.*, **3(3)**, 191–5.

Demoll, R. (1918). *Der Flug der Insekten und der Vögel.* Gustav Fischer, Jena.

Deshpande, S.B. (1984). The thoracic functional myology of odonates—a study of the patterns as found in *Lestes elata*, Hagen (Zygoptera) and *Anax immaculifrons*, Rambur (Anisoptera). *J. Anim. Morphol. Physiol.*, **31(1–2)**, 67–77.

Dicke, R.J. and Howe, H.M. (1978). A revision of morphological terminology. *Ann. Ent. Soc. Amer.*, **71(2)**, 228–38.

Dumont, H. (1964). Note on a migration of the dragonfly *Libellula quadrimaculata* L. in the North of France. *Bull. et Ann. Soc. roy. ent. Belg.*, **100(13)**, 177–81.

Duncan, W.J., Thom, A.S., and Young, A.D. (1970). *Mechanics of fluids.* Edward Arnold, London.

Dworakowska, I. (1988). Main veins of the wings of Auchenorrhyncha. *Ent. Abhandlungen*, **52(3)**, 63–108.

Edmunds, G.F. and Traver, J.R. (1954). The flight mechanics and evolution of the wings of Ephemeroptera, with notes on the archetype insect wing. *J. Wash. Acad. Sci.*, **44(12)**, 390–400.

Ellington, C.P. (1980). Vortices and hovering flight. In *Instationäre Effekte an schwingenden Tierflügeln* (ed. W. Nachtigall), pp.64–101. Franz Steiner, Wiesbaden.

Ellington, C.P. (1984). The aerodynamics of hovering insect flight. *Phil. Trans. Roy. Soc. London*, **305(1122)**, 1–181.

Emelyanov, A.F. (1977). Homologies in wing structure between cicadas and primitive Polyneoptera. *Trudy VEO*, **58**, 3–48. (In Russian.)

Emelyanov, A.F. (1987). The phylogeny of Homoptera, Cicadina based on comparative data. *Trudy VEO*, **69**, 19–109. (In Russian.)

Ennos, A.R. (1987). A comparative study of the flight mechanism of Diptera. *J. exp. Biol.*, **127**, 355–72.

Ennos, A.R. and Wootton, R.J. (1989). Functional wing morphology and aerodynamics of *Panorpa germanica* (Insecta: Mecoptera). *J. exp. Biol.*, **143**, 267–84.

Esch, H., Nachtigall, W., and Kogge, S.N. (1975). Correlations between aerodynamic output, electrical activity in the indirect flight muscles and wing positions of bees flying in a servo-mechanically controlled wind tunnel. *J. Comp. Physiol.*, **A100(2)**, 147–59.

Evans, J.W. (1941). The morphology of *Tettigarcta tomentosa* White (Homoptera, Cicadidae). *Pap. Proc. Roy. Soc. Tasmania*, **1940**, 35–49.

Flower, J.W. (1964). On the origin of flight in insects. *J. Insect Physiol.*, **10(1)**, 81–8.

Forbes, W.T.M. (1943). The origin of wings and venational types in insects. *Amer. Midl. Nat.*, **29(2)**, 381–405.

Fung, Y.C. (1969). *An introduction to the theory of aeroelasticity.* Dover, New York.

Gatter, W. (1981). Anpassungen von Wanderinsekten an die tagliche Drehung des Windes. *Jahresh. Ges. Naturk. Württemberg*, **136**, 191–202.

Gewecke, M. (1970). Antennae: another wind-sensitive receptor in locusts. *Nature*, **225(5239)**, 1263–4.

Gewecke, M. and Niehaus, M. (1981). Flight and flight control by the antennae in the small Tortoiseshell (*Aglais urticae* L., Lepidoptera). I. Flight balance experiments. *J. Comp. Physiol.*, **A145(2)**, 270–86.

George, J.C. and Bhakthan N.M.G. (1960). A study on the fibre diameter and certain enzyme concentrations in the flight muscles of some butterflies. *J. exp. Biol.*, **37(2)**, 308–15.

Gibo, D.L. and Pallett, M.J. (1979). Soaring flight of monarch butterflies, *Danaus plexippus* (Lepidoptera: Danaidae), during the late summer migration in Southern Ontario. *Canad. J. Zool.*, **57(7)**, 1393–401.

Gibson, G.A.P. (1986). Mesothoracic skeletomusculature and mechanics of flight and jumping in Eupelminae (Hymenoptera, Chalcidoidea: Eupelmidae). *Canad. Entomol.*, **118(7)**, 691–728.

Golubev, V.V. (1957). *Works on aerodynamics.* Gos. isd. tech.-teor. literatury, Moscow, Leningrad. (In Russian.)

Gorelov, D.N. (1976). On the effectivenes of the flapping wing as a propulsive organ. *Bionika*, **10**, 49–53. (In Russian.)

Govind, C.K. (1972). Differential activity in the coxo-subalar muscle during directional flight in the milkweed bug *Oncopeltus*. *Canad. J. Zool.*, **50(6)**, 901–5.

Grebeshov, E.P. and Sagoyan, O.A. (1976). Hydrodynamic features of the oscillating wing, working as a supporting gliding plane and a propulsive organ. *Trudy TZAGI*, **1725**, 3–30. (In Russian.)

Griffith, S.N. (1972). Migrations of the day-flying moth *Urania* in Central and South America. *Carib. J. Sci.*, **12(1–2)**, 45–58.

Grodnitsky, D.L. and Kozlov, M.V. (1985). Functional morphology of the wing apparatus and flight characteristics of primitive moths (Lepidoptera: Micropterigidae, Eriocraniidae). *Zool. Zh.*, **64(11)**, 1661–71. (In Russian.)

Hatch, G. (1966). Structure and mechanics of the dragonfly pterothorax. *Ann. ent. Soc. Am.*, **59**, 702–14.

Helfert, M.R. (1972). Migratory behavior of the snout butterfly, *Libytheana bachmanii* (Strecker). *Ent. News*, **83(2)**, 49–52.

Hepburn, H.R. (1970). The skeleto-muscular system of Mecoptera: the thorax. *Univ. Kansas Science bulletin*, **48(21)**, 801–44.

Hertel, H. (1966). *Structure—form—movement*. Reinhold, New York.

Hlavač, T.F. (1974). *Merope tuber* (Mecoptera): a wing-body interlocking mechanism. *Psyche*, **81(2)**, 303–6.

Holst, E. (1943). Untersuchungen über Flugbiophysik. I. Messungen zur Aerodynamik kleiner schwingender Flügel. *Biol. Zbl.*, **63(7–8)**, 289–326.

Horridge, G.A. (1956). The flight of very small insects. *Nature*, **178**, 45–6.

Hynes, H.B.N. (1974). Observations on the adults and eggs of Australian Plecoptera. *Austral. J. Zool. Suppl. Ser.*, **29**, 37–52.

Ivanov, V.D. (1981). Wing folding in living insects. *Vestnik Leningr. Univ.* (Biology), **15**, 101–3. (In Russian.)

Ivanov, V.D. (1985). A comparative study of wing kinematics in caddisflies (Trichoptera). *Ent. obozr.*, **64(2)**, 273–84. (In Russian.)

Ivanov, V.D. (1990). A comparative analysis of flight aerodynamics of caddisflies (Insecta: Trichoptera). *Zool. Zh.*, **69(2)**, 46–60. (In Russian.)

Ivanov, V.D. and Kozlov, M.V. (1987). A comparative analysis of pterothoracic musculature of caddis-flies (Insecta, Trichoptera). *Zool. Zh.*, **66(10)**, 1484–98. (In Russian.)

Jensen, M. (1956). Biology and physics of locust flight. III. The aerodynamics of locust flight. *Phil. Trans. Roy. Soc. London, B*, **239(667)**, 511–52.

Johnson, C.G. (1969). *Migration and dispersal of insects by flight*. Methuen, London.

Kammer, A.E. (1971). The motor output during turning flight in a hawkmoth, *Manduca sexta*. *J. Insect Physiol.*, **17(6)**, 6, 1073–86.

Kármán, T. von and Burgers J.M. (1935). General aerodynamic theory—perfect fluids. In *Aerodynamic theory* (ed. W. Durand), Vol.2, Div.E. Springer, Berlin.

Kluge, N.J. (1989). A question of the homology of the tracheal gills and paranotal processi of mayfly larvae and the wings of insects with reference to the taxonomy and phylogeny of the order Ephemeroptera. *Lecture series in memory of N. A. Kholodkovsky*, **41**, 42–77. (In Russian.)

Kokshaysky, N. V. (1974). *An essay on biological aero- and hydrodynamics (flight and swimming of animals)*. Nauka, Moscow. (In Russian.)

Kokshaysky, N. V. (1979). Tracing the wake of a flying bird. *Nature*, **279** (5709), 146–8.

Kokshaysky, N.V. (1982). The role of Russian science in bird flight studies. *Zool. Zh.*, **61(7)**, 971–87. (In Russian.)

Kozlov, M.V. (1986). Pterothoracical muscles of the primitive moths (Lepidoptera, Micropterigidae—Tischeriidae). *Vestnik zool.*, **1**, 60–71. (In Russian.)

Kozlov, M. V., Ivanov, V.D., and Grodnitsky, D.L. (1986). Evolution of the wing apparatus and wing kinematics in Lepidoptera. *Uspehi sovr. biology*, **101(2)**, 291–305. (In Russian.)

Kukalová, J. (1958). Paoliidae Handlirsch (Insecta—Protorthoptera) aus dem Oberschlesischen Steinkohlenbedecken. *Geologie*, **7(7)**, 935–59.

Kukalová-Peck, J. (1972). Unusual structures in the Palaeozoic insect orders Megasecoptera and Palaeodictyoptera, with a description of a new family. *Psyche*, **79(3)**, 243–68.

Kukalová-Peck, J. (1974). Pteralia of the Palaeozoic insect orders Palaeodictyoptera, Megasecoptera, and Diaphanopterodea (Paleoptera). *Psyche*, **81 (3–4)**, 416–29.

Kukalová-Peck, J. (1978). Origin and evolution of insect wings and their relation to metamorphosis, as documented by the fossil record. *J. Morphol.*, **156(1)**, 53–126.

Kukalová-Peck, J. (1983). Origin of the insect wing

and wing articulation from the arthropodan leg. *Canad. J. Zool.*, **61(7)**, 1618–69.

Kukalová-Peck, J. (1985). Ephemeroid wing venation based upon new gigantic Carboniferous mayflies and basic morphology, phylogeny, and metamorphosis of pterygote insects (Insecta, Ephemerida). *Canad. J. Zool.*, **63**, 933–55.

Kukalová-Peck, J. (1987). New Carboniferous Diplura, Monura, and Thysanura, the hexapod ground plan, and the role of thoracic lobes in the origin of wings (Insecta). *Canad. J. Zool.*, **65**, 2327–45.

Kukalová-Peck, J. and Brauckmann, C. (1990). Wing folding in pterygote insects, and the oldest Diaphanopterodea from the early Late Carboniferous of West Germany. *Canad. J. Zool.*, **68**, 1104–11.

Kukalová-Peck, J. and Richardson, E.S. (1983). New Homoiopteridae (Insecta: Paleodictyoptera) with wing articulation from Upper Carboniferous strata of Mazon Creek, Illinois. *Canad. J. Zool.*, **61(7)**, 1670–87.

Kutsch, W. (1974). The development of the flight pattern in locusts. *Exp. Anal. Insect Behaviour.* 149–58.

Labedzki, A. (1982). Masowy pojaw wazki *Libellula quadrimaculata* L. w Swietokrzyskim Parku Narodowym. *Parki Nar. i reserw. przyr.*, **3(1)**, 19–22.

La Greca, M. (1947). Morfologia funzionale dell'articolazione alare degli Ortotteri. *Arch. Zool. Ital.*, **32**, 271–327.

Larsen, O. (1949). Die Orstbewegungen von *Ranatra linearis* L. *Lunds Univ. Årsskr.*, **45(6)**, 1–83.

Magnan, A. (1934). *La locomotion chez les animaux. I. Le vol des insectes.* Hermann et cie, Paris.

Marey, C. (1869). Mémoire sur le vol des insectes et des oiseaux. *Ann. Sci. Nat. Zool.*, **5(12)**, 49–150.

Martin, L.J. and Carpenter, P.W. (1977). Flow-visualization experiments on butterflies in simulated gliding flight. In *The physiology of movement; biomechanics*, (ed. W. Nachtigall), pp.307–15. Fischer, Stuttgart.

Martynov, A.V. (1924). L'évolution de deux formes d'ailes différentes chez les insectes. *Russk. Zool. Zh.*, **4**, 155–85. (In Russian, with French summary.)

Martynov, A.V. (1928). Permian fossil insects of north-east Europe. *Trudy goel. Muz.*, **4**, 1–118. (In Russian.)

Martynov, A.V. (1938). Studies on the geological history and phylogeny of the orders of insects (Pterygota). *Trudy paleont. Inst.*, **7(4)**, 1–149. (In Russian, with French summary.)

Matsuda, R. (1970). Morphology and evolution of the insect thorax. *Mem. Entomol. Soc. Canad.*, **76**, 1–431.

Matsuda, R. (1981). The origin of insect wings (Arthropoda: Insecta). *Int. J. Insect Morph. Embr.*, **10(5/6)**, 387–98.

Maxworthy, T. (1981). The fluid dynamics of insect flight. *Ann. Rev. Fluid Mech.*, **13**, 329–50.

Mertens, H. (1923). Biologische und morphologische Untersuchungen an Plekopteren. *Archiv Naturg. Abteilung*, **A(2)**, 1–38.

Mickoleit, G. (1966). Zur Kenntnis einer neuen Spezialhomologie (Synapomorphie) der Panorpoidea. *Zool. Jb. Anat.*, **83**, 483–97.

Mises, R. von (1959). *Theory of flight.* Dover, New York.

Miyan, J.A. and Ewing, A.W. (1985). How Diptera move their wings: a re-examination of the wing base articulation and muscle systems concerned with flight. *Phil. Trans. Roy. Soc. London. B*, **311(1150)**, 271–302.

Nachtigall, W. (1966). Die Kinematik der Schlagflügelbewegungen von Dipteren. Methodische und analytische Grundlagen zur Biophysik des Insektenflugs. *Z. vergl. Physiol.*, **52(2)**, 155–211.

Nachtigall, W. (1967). Aerodynamische Messungen am Tragflügelesystem segelender Schmetterlinge. *Z. vergl. Physiol.*, **54(2)**, 210–31.

Nachtigall, W. (1969). Über den Start des fibrillären Flugmotors dei calliphoriden Dipteren. *Zool. Anz. Suppl.*, **32**, 444–8.

Nachtigall, W. (1974). *Insects in flight. A glimpse behind the scenes in biophysical research.* McGraw-Hill, New York.

Nachtigall, W. (1977). Die aerodynamische Polare des Tipula-Flügels und eine Einrichtung zur halbautomatischen Polarenaufnahme. In *The physiology of movement; biomechanics*, (ed. W. Nachtigall), p.347–52. Fischer, Stuttgart.

Nachtigall, W. (1978). Der Startsprung der Stubenfliege *Musca domestica* (Diptera: Muscidae). *Entomol. German.*, **4(3–4)**, 368–73.

Nachtigall, W. (1979). Schiebeflug bei der Schmeissfliege *Calliphora erythrocephala* (Diptera: Calliphoridae). *Entomol. Gen.*, **5(3)**, 255–65.

Nachtigall, W. (1980). Rasche Bewegungsänderungen bei der Flügelschwingung von Fliegen und ihre mögliche Bedeutung für instationäre Luftkrafterzeugung. In *Instationäre Effekte an schwingenden Tierflügeln*, (ed. W. Nachtigall), pp.115–29. Franz Steiner, Wiesbaden.

Nachtigall, W. (1981). Hydromechanics and biology. *Biophys. Struct. Mech.*, **8**, 1–22.

Nachtigall, W. and Roth, W. (1983). Correlations between stationary measurable parameters of wing movement and aerodynamic force production in the Blowfly (*Calliphora vicina* R.-D.). *J. Comp. Physiol.*, **A150(2)**, 251–60.

Nachtigall, W. and Wilson, D. (1967). Neuro-muscular control of dipteran flight. *J. exp. Biol.*, **47**, 77–97.

Neville, A.C. (1960). Aspects of flight mechanics in anisopterous dragonflies. *J. exp. Biol.*, **37**, 631–56.

Newman, B.G., Savage, S.B., and Schouella, D. (1977). Model tests on a wing section of an *Aeschna* dragonfly. In *Scale effects in animal locomotion*, (ed. T.J. Pedley), pp.445–77. Academic Press, London.

Newman, D.J.S. and Wootton, R.J. (1988). The role of the fulcroalar muscle in dragonfly flight. *Odonatologica*, **17(4)**, 401–8.

Norberg, R.Å. (1972a) Flight characteristics of two moths, *Alucita pentadactyla* L. and *Orneodes hexadactyla* L. (Microlepidoptera). *Zool. Scr.*, **1**, 241–6.

Norberg, R.Å. (1972b). The pterostigma of insect wings as inertial regulator of wing pitch. *J. Comp. Physiol..*, **81(1)**, 9–22.

Norberg, R.Å. (1975). Hovering flight of the dragonfly *Aeschna juncea* L., kinematics and aerodynamics. In *Swimming and flying in nature*, (ed. T.Y. Wu, C.J.Brokaw, and C. Brennen), pp. 763–81. Plenum Press, New York.

Pannycuick, C. J. (1972). Soaring behavior and performance of some East African birds, observed from a motor-glider. *Ibis*, **114**, 178–218.

Pfau, H.K. (1986). Untersuchungen zur Konstruktion, Funktion und Evolution des Flugapparates der Libellen (Insecta, Odonata). *Tijdschrift voor Entomologie*, **129(3)**, 35–123.

Pfau, H.K. (1987). Critical comments on a 'novel mechanical model of dipteran flight' (Miyan, Ewing, 1985). *J. exp. Biol.*, **128**, 463–8.

Pfau, H.K. and Honomiche, K. (1979). Die campaniformen sensillen des Flügels von *Cetonia aurata* L. und *Geotrupes silvaticus* Pauz. (Insecta, Coleoptera) in Bezeihung zur Flügelmechanik und Flügfunktion. *Zool. Jb. Anat. Ontog. Tiere*, **102**, 583–613.

Pfau, H.K. and Nachtigall, W. (1981). Der Vorderflügel grosser Heuschrecken als Luftkrafterzeuger. II. Zusammenspiel von Muskeln und Gelenkmechanik dei der Einstellung der Flügelgeometrie. *J. Comp. Physiol.*, **142**, 135–40.

Polonsky, Ya.E. (1950). Some questions about wing flapping. *Inzh. sbornik*, **8**, 49–60. (In Russian.)

Ponomarenko, A.G. (1980). In *Historical development of the Class Insecta. Trudy paleont. Inst.*, **175**, p.46. Nauka, Moscow. (In Russian.)

Prandtl, L. and Tietjens, O.G. (1957). *Applied hydro- and aeromechanics*. Dover, New York.

Pringle, J.W.S. (1957). *Insect flight*. Cambridge University Press.

Pringle, J.W.S. (1968). Comparative physiology of the flight motor. *Adv. Insect Physiol.*, **5**, 163–223.

Pritykina, L.N. (1989). Palaeontology and evolution of dragonflies. In *Fauna and ecology of dragonflies* (ed. V.G.Mordkovich), pp. 43–58. Nauka, Novosibirsk.

Ragge, D.R. (1972). An unusual case of mass migration by flight in *Gryllus bimaculatus* Degeer (Orthoptera: Gryllidae). *Bull. Inst. Fondam. Afr. noire*, **A34(4)**, 869–78.

Rainey, R.C. (1985). Insect flight: new facts—and old fantasies? In *Insect locomotion*. Proceedings of symposium 4.5, 17th International Congress of Entomology, (ed. M. Gewecke and G. Wendler), pp.241–4. Paul Parey, Berlin, Hamburg.

Rasnitsyn, A.P. (1969). The origin and evolution of the lower Hymenoptera. *Trudy paleont. Inst.*, **123**, 1–196. (In Russian.)

Rasnitsyn, A.P. (1976). On the early evolution of insects and the origin of Pterygota. *Zh. obsch. biol.*, **37**, 543–55. (In Russian.)

Rasnitsyn, A.P. (1980a). Origin and evolution of the Hymenoptera. *Trudy paleont. Inst.*, **174**, 1–192. (In Russian.)

Rasnitsyn, A.P. (1980b). Subclass Scarabaeona Laicharting, 1781. In *Historical development of the Class Insecta. Trudy paleont. Inst.*, **175**, p.24–30. Nauka, Moscow. (In Russian.)

Rasnitsyn, A.P. (1980c). Superorder Blattidea

Latreille, 1810. In *Historical development of the Class Insecta. Trudy paleont. Inst.*, **175**, p.136–48. Nauka, Moscow. (In Russian.)

Rasnitsyn, A.P. (1980*d*). Cohort Cimiciformes Laicharting, 1781. In *Historical development of the Class Insecta. Trudy paleont. Inst.*, **175**, p.36–41. Nauka, Moscow. (In Russian.)

Rasnitsyn, A.P. (1981). A modified paranotal theory of insect wing origin. *J. Morphol.*, **168**, 331–8.

Rayner, J.M.V. (1979). A new approach to animal flight mechanics. *J. exp. Biol.*, **80**, 17–54.

Rayner, J.M.V. (1980). Vorticity and animal flight. In *Aspects of animal movement* (ed. H.Y.Elder and E.R. Trueman), p.177–99. Cambridge University Press.

Richards, O.W. (1956). *Hymenoptera: Introduction and key to families.* Royal Entomological Society of London.

Riek, E.F. and Kukalová-Peck, J. (1984). A new interpretation of dragonfly wing venation based upon Early Upper Carboniferous fossils from Argentina (Insecta: Odonatoidea) and basic character states in pterygote wings. *Canad. J. Zool.*, **62(6)**, 1150–66.

Riley, J.R., Reynolds D.R., and Farmery, M.J. (1983). Observations of the flight behaviour of the armyworm moth, *Spodoptera exepta*, at an emergence site using radar and infra-red optical techniques. *Ecol. Entomol.*, **8(4)**, 395–418.

Robertson, R.M., Pearson, K.G., and Reichert, H. (1982). Flight interneurones in the locust and the origin of insect wings. *Science* (Washington, D.C.), **217**, 177–9.

Rohdendorf, B.B. (1949). Evolution and classification of flight apparatuses in insects. *Trudy paleont. Inst.*, **16**. (In Russian.)

Rudolph, R. (1976). Die aerodynamischen Eigenschaften von *Calopteryx splendens* (Harris) (Zygoptera: Calopterygidae). *Odonatologica*, **5(4)**, 383–6.

Rüppell, G. (1985). Kinematic and behavioural aspects of flight of the male banded agrion, *Calopteryx* (*Agrion*) *splendens* L. In *Insect locomotion.* Proceedings of Symposium 4.5, 17th International Congress of Entomology, (ed. M. Gewecke and G. Wendler), pp.195–204. Paul Parey, Berlin, Hamburg.

Russev, B. (1973). Kompensationflug bei der Ordnung Ephemeroptera. *Proc. 1st Int. Conf. on Ephemeroptera*, (ed. W.L. Peters and J.G. Peters) pp.132–42. Brill, Leiden.

Ryazanova, G.I. (1965). Body weight to wing area ratio in dragonflies. *Zool. Zh.*, **44(a)**, 1357–62. (In Russian.)

Ryazanova, G.I. (1966). Comparative characteristics of dragonfly flight. *Zh. obsch. biol.*, **27(3)**, 349–59. (In Russian.)

Savinov, A.B. (1983). Morphology of the thoracic skeleton in the nymph and imago of the green leafhopper *Cicadella viridis* L. (Homoptera, Cicadellidae). *Ent. obozr.*, **62(4)**, 673–89. (In Russian.)

Savinov, A.B. (1986). Morphology of the nymphal and imaginal skeleton of the thorax in the froghopper *Aphrophora salicina* Goeze (Homoptera, Aphrophoridae). *Ent. obozr.*, **65(1)**, 43–58. (In Russian.)

Schlichting, H. (1974). *Grenzschicht—Theorie.* Verlag G. Braun, Karlsruhe.

Schneider, P. (1980). Beiträge zur Flugbiologie der Käfer. 5. Kinematik der Alae und vertikale Richtungsanderung. *Zool. Anz.*, **3/4**, 188–98.

Schneider, P. (1982). Untersuchungen zur Steuerung des Flugmotors beim Maikafer. In *Insect flight* Biona Report 1, (ed. W. Nachtigall), p. 121–33 Fischer, Stuttgart.

Schouest, L.P., Anderson, M., and Miller, T.A. (1986). The ultrastructure and physiology of the tergotrochanteral depressor muscle of the housefly *Musca domestica. J. Exp. Zool.*, **239(2)**, 147–58.

Schrott, A. (1986). Vergleichende Morphologie und Ultrastruktur des Cenchrus—Dornenfeld apparates bei Pflanzenwespen (Insecta: Hymenoptera, Symphyta). *Ber. naturwiss.-med. Ver. Innsbruck*, **73**, 159–68.

Scott, J. (1975). Movements of *Euchloe ausonides* (Pieridae). *J. Lepidopter. Soc.*, **29(1)**, 24–31.

Sharov, A.G. (1966). *Basic arthropodan stock with special reference to insects.* Pergamon Press Oxford.

Sharov, A.G. (1968). The phylogeny of the orthopteroid insects. *Trudy paleont. Inst.*, **118**, 1–217 (In Russian.)

Sharplin, J. (1963*a*). Wing base structure in Lepidoptera. I. Fore wing base. *Canad. Entomol.* **95(10)**, 1024–50.

Sharplin, J. (1963*b*). Wing base structure in Lepidoptera. II. Hind wing base. *Canad. Entomol.* **95(11)**, 1121–45.

Shvanvich, B.N. (1949). *General course of entomology.* Soviet Nauka, Moscow. (In Russian.)

Sinichenkova, N.D. (1987). Historical development of stoneflies. *Trudy paleont. Inst.*, **221**, 1–143. (In Russian.)

Snodgrass, R.E. (1927). Morphology and mechanism of insect thorax. *Smithsonian Misc. Coll.*, **80**, 1.

Snodgrass, R.E. (1929). The thoracic mechanism of a grasshopper and its antecendents. *Smithsonian Misc. Coll.*, **82**, 2.

Snodgrass, R.E. (1930). How insects fly. *Smithsonian Rept.*, 1929, 383–421.

Snodgrass, R.E. (1935). *Principles of insect morphology.* McGraw-Hill, New York, London.

Sotavalta, O. (1947). The flight tone (wing-stroke frequency) of insects. *Acta Ent. Fenn.*, **4**, 1–117.

Sotavalta, O. (1952). The essential factor regulating the wing stroke frequency of insects in wing mutilation and loading experiments and in experiments at subatmospheric pressure. *Ann (bot.-zool.) Soc. zool. -bot. fenn. Vanamo (Zool.)*, **15**, 1–67.

Spedding, G.R. (1986). The wake of a jackdaw (*Corvus monedula*) in slow flight. *J. exp. Biol.*, **125**, 287–307.

Spedding, G.R., Rayner, J.M.V., and Pennycuick, C.J. (1984). Momentum and energy in the wake of a pigeon (*Columba livia*) in slow flight. *J. exp. Biol.*, **111**, 81–102.

Spüler, M. and Heide, G. (1978). Simultaneous recordings of torque, thrust and muscle spikes from the fly *Musca domestica* during optomotor responses. *Z. Naturforsch.*, **3(5–6)**, 455–7.

Stellwaag, F. (1916). Wie steuern die Insecten während des Fluges? *Biol. Zentralbl.*, **36(1)**, 30–44.

Svidersky, V.L. (1973). *Neurophysiology of insect flight.* Nauka, Leningrad. (In Russian.)

Tannert, W. (1958). Die Flügelgekenkung bei Odonaten. *Dt. ent. Z. (N.F.)*, **5**, 394–455.

Taylor, L.H. (1918). The thoracic sclerites of Hemiptera and Heteroptera. *Annals ent. Soc. Am.*, **11(3)**, 225–54.

Taylor, L.R. (1974). Insect migration, flight periodicity and the boundary layer. *J. Anim. Ecol.*, **43(1)**, 225–38.

Termier, M. (1970). Conséquences des capacités d'accélération en vol des diptères syrphides et bombylides. *C. R. Acad. sci.* D **271(25)**, 2361–3.

Tiegs, O.W. (1955). The flight muscles of insects—their anatomy and physiology, with some observations on the structure of striated muscle in general. *Phil. Trans. Roy. Soc. Lond. B*, **238**, 221–359.

Tietze, F. (1963). Untersuchungen über die Beziehungen zwischen Flügelreduktion und Ausbildung des Metathorax bei Carabiden unter besonderer Burücksichtigung der Flugmuskulatur. *Beitr. Ent.*, **13(1–3)**, 88–167.

Vinogradov, R.I. (1959). The dependence of the reversed vortex street on Strouhal number. *Izvestia vysch. uch. zaved., Aviatz. Tehn.*, **4**, 14–25. (In Russian.)

Vogel, S. (1967a). Flight in *Drosophila*. II. Variations in stroke parameters and wing contour. *J. exp. Biol.*, **46**, 383–92.

Vogel, S. (1967b). Flight in *Drosophila*. III. Aerodynamic characteristics of fly wings and wing models. *J. exp. Biol.*, **46**, 431–43.

Voisin, J.-F. (1976). De l'utilisation de ses organes de vol par la femelle de *Chorthippus longicornis* (Orth. Acrididae). *Entomologiste*, **32(3)**, 135–6.

Wagner, H. (1985). Aspects of the free flight behaviour of houseflies (*Musca domestica*). In *Insect locomotion*. Proceedings of Symposium 4.5, 17th International Congress of Entomology, (ed. M. Gewecke and G. Wendler), pp.223–32. Paul Parey, Berlin, Hamburg.

Wagner, S. (1980). Einfluss der instationären Strömung auf die Aerodynamik des Hubschrauberrotors. In *Instationäre Effekte an schwingenden Tierflügeln*, (ed. W. Nachtigall), p.35–59. Franz Steiner, Wiesbaden.

Weber, H. (1929). Kopf und Thorax von *Psylla mali* Schmidb. *Z. Morph. Ökol. Tiere*, **14**, 60–165.

Weis-Fogh, T. (1956). Biology and physics of locust flight. II. Flight performance of the desert locust (*Schistocerca gregaria*). *Phil. Trans. Roy. Soc. Lond. B*, **239(667)**, 459–510.

Weis-Fogh, T. (1972). Energetics of hovering flight in humming birds and in *Drosophila*. *J. exp. Biol.*, **56**, 79–104.

Weis-Fogh, T. (1973). Quick estimates of flight fitness in hovering animals, including novel mechanisms for lift production. *J. exp. Biol.*, **59**, 169–230.

Weis-Fogh, T. (1975). Flapping flight and power in birds and insects, conventional and novel mechanisms. In *Swimming and flying in Nature*, (ed. T.Y. Wu, C.J. Brokaw, and C. Brennen), pp. 729–62. Plenum Press, New York.

Weis-Fogh, T. and Jensen, M. (1956). Biology and physics of locust flight. I. Basic principles of insect flight: a critical review. *Phil. Trans. Roy. Soc. Lond. B*, **239(667)**, 415–58.

Whitten, J.M. (1962). Homology and development of insect wing tracheae. *Ann. Ent. Soc. Am.*, **55(3)**, 288–95.

Wilson, D.M. (1968a). The flight-control system of the locust. *Sci. Amer.*, **218(5)**, 83–90.

Wilson, D.M. (1968b). The nervous control of insect flight and related behaviour. *Adv. Insect Physiol.*, **5**, 289–338.

Wilson, D.M. and Weis-Fogh, T. (1962). Pattern activity of coordinated motor units, studied in flying locusts. *J. exp. Biol.*, **39**, 643–67.

Wootton, R.J. (1976). The fossil record and insect flight. In *Insect flight*, (ed. R.C. Rainey), pp. 235–54. Blackwell, London. Oxford.

Wootton, R.J. (1979). Function, homology and terminology in insect wings. *Syst. Entomol.*, **4(1)**, 81–93.

Wootton, R.J. (1981). Support and deformability in insect wings. *J. Zool.*, **193**, 447–68.

Wootton, R.J. and Betts, C.R. (1986). Homology and function in the wings of Heteroptera. *Syst. Entomol.*, **11**, 389–400.

Wootton, R.J. and Ennos, A.R. (1989). The implications of function on the origin and homologies of the dipterous wing. *Syst. Entomol.*, **14**, 507–20.

Wootton, R.J. and Newman, D.J.S. (1979). White fly have the highest contraction frequences yet recorded in non-fibrillar flight muscles. *Nature* **280(5721)**, 402–3.

Zalessky, Yu.M. (1949). Wing origin and the appearance of flight in insects constrained by environmental conditions. *Uspehi sovr. biol* **28(3)**, 400–14. (In Russian.)

Zalessky, Yu.M. (1955). The current state of insect flight studies. *Uspehi sovr. biol.*, **39(3)**, 308–27. (In Russian.)

Zarnack, W. (1972). Flugbiophysik der Wanderheuschrecke *(Locusta migratoria* L.). I. Die Bewegungen der Vorderflügel. *J. Comp. Physiol.*, **78(4)**, 356–95.

Taxonomic index

Numbers in italics refer to figures on the given pages

Subject index

Numbers in italics refer to figures on the given pages